National Research Council. Committee on Biologic Effects of Atmospheric Pollutants.

Medical and Biologic Effects of Environmental Pollutants

CHROMIUM

Committee on
Biologic Effects of
Atmospheric Pollutants

DIVISION OF MEDICAL SCIENCES
NATIONAL RESEARCH COUNCIL

NATIONAL ACADEMY
OF SCIENCES

WASHINGTON, D.C.   1974

Other volumes in the Medical and Biologic Effects of Environmental Pollutants series (formerly named Biologic Effects of Atmospheric Pollutants):

ASBESTOS (ISBN 0-309-01927-3)
FLUORIDES (ISBN 0-309-01922-2)
LEAD (ISBN 0-309-01941-9)
PARTICULATE POLYCYCLIC ORGANIC MATTER (ISBN 0-309-02027-1)
MANGANESE (ISBN 0-309-02143-X)
VANADIUM (ISBN 0-309-02218-5)

NOTICE: The project that is the subject of this report was approved by the Governing Board of the National Research Council, acting in behalf of the National Academy of Sciences. Such approval reflects the Board's judgment that the project is of national importance and appropriate with respect to both the purposes and the resources of the National Research Council.

The members of the committee selected to undertake this project and prepare this report were chosen for recognized scholarly competence and with due consideration for the balance of disciplines appropriate to the project. Responsibility for the detailed aspects of this report rests with that committee.

Each report issuing from a study committee of the National Research Council is reviewed by an independent group of qualified persons according to procedures established and monitored by the Report Review Committee of the National Academy of Sciences. Distribution of the report is approved, by the President of the Academy, upon satisfactory completion of the review process.

The work on which this publication is based was performed pursuant to Contract No. 68-02-0542 with the Environmental Protection Agency.

Library of Congress Cataloging in Publication Data

National Research Council. Committee on Biologic
Effects of Atmospheric Pollutants.
Chromium.

(Its medical and biologic effects)
Bibliography: p.
1. Chromium–Toxicology. 2. Environmental health.
I. Title. [DNLM: 1. Chromium. 2. Environmental
pollution. QV290 N277c 1974]
RA1231.C5N47 1974 574.2'4 74-3493
ISBN 0-309-02217-7

Available from
Printing and Publishing Office, National Academy of Sciences
2101 Constitution Avenue, Washington, D.C. 20418

Printed in the United States of America

PANEL ON CHROMIUM

ANNA M. BAETJER, Department of Environmental Medicine, School of Hygiene and Public Health, The Johns Hopkins University, Baltimore, Maryland, *Chairman*

DONALD J. BIRMINGHAM, Department of Dermatology and Syphilology, Wayne State University Medical School, Detroit, Michigan

PHILIP E. ENTERLINE, Department of Biostatistics, Graduate School of Public Health, University of Pittsburgh, Pittsburgh, Pennsylvania

WALTER MERTZ, Vitamin and Mineral Nutrition Laboratory, Human Nutrition Research Division, Agricultural Research Service, U.S. Department of Agriculture, Beltsville, Maryland

JAMES O. PIERCE, II, Environmental Trace Substances Center, University of Missouri, Columbia, Missouri

P. LESLEY BIDSTRUP, Sloane Terrace Mansions, London, England, *Consultant*

W. CLARK COOPER, Tabershaw-Cooper Associates, Berkeley, California, *Associate Editor*

DAVID M. ANDERSON, Environmental Quality Control Division, Industrial Relations Department, Bethlehem Steel Corporation, Bethlehem, Pennsylvania, *BEAP Committee Liaison*

JOHN REDMOND, JR., Division of Medical Sciences, National Research Council, Washington, D.C., *Staff Officer*

COMMITTEE ON BIOLOGIC EFFECTS OF ATMOSPHERIC POLLUTANTS

HERSCHEL E. GRIFFIN, Graduate School of Public Health, University of Pittsburgh, Pittsburgh, Pennsylvania, *Chairman*

DAVID M. ANDERSON, Environmental Quality Control Division, Industrial Relations Department, Bethlehem Steel Corporation, Bethlehem, Pennsylvania

VINTON W. BACON, Department of Civil Engineering, College of Applied Science and Engineering, University of Wisconsin, Milwaukee, Wisconsin

ANNA M. BAETJER, Department of Environmental Medicine, School of Hygiene and Public Health, The Johns Hopkins University, Baltimore, Maryland

RICHARD U. BYERRUM, College of Natural Science, Michigan State University, East Lansing, Michigan

W. CLARK COOPER, Tabershaw-Cooper Associates, Berkeley, California

SAMUEL A. GUNN, Department of Pathology, University of Miami School of Medicine, Miami, Florida

ROBERT I. HENKIN, National Heart and Lung Institute, National Institutes of Health, Bethesda, Maryland

SAMUEL P. HICKS, Department of Pathology, University of Michigan Medical Center, Ann Arbor, Michigan

IAN T. T. HIGGINS, School of Public Health, University of Michigan, Ann Arbor, Michigan

JAN LIEBEN, American Viscose Division, FMC Corporation, Philadelphia, Pennsylvania

JAMES N. PITTS, JR., Statewide Air Pollution Control Center, University of California, Riverside, California

I. HERBERT SCHEINBERG, Albert Einstein College of Medicine, Bronx, New York

RALPH G. SMITH, Department of Environmental and Industrial Health, School of Public Health, University of Michigan, Ann Arbor, Michigan

GORDON J. STOPPS, Environmental Health Branch, Ontario Department of Health, Toronto, Ontario, Canada

F. WILLIAM SUNDERMAN, JR., Department of Laboratory Medicine, University of Connecticut School of Medicine, Farmington, Connecticut

BENJAMIN L. VAN DUUREN, Institute of Environmental Medicine, New York University Medical Center, New York, New York

JAROSLAV J. VOSTAL, Department of Pharmacology and Toxicology, University of Rochester Medical Center, Rochester, New York

BERNARD WEISS, Department of Radiation Biology and Biophysics, University of Rochester Medical Center, Rochester, New York

T. D. BOAZ, JR., Division of Medical Sciences, National Research Council, Washington, D.C., *Executive Director*

Acknowledgments

This document was written by the Panel on Chromium under the chairmanship of Dr. Anna M. Baetjer. Each section was prepared initially by a member of the Panel, but the total document was reviewed by the entire Panel and represents the combined effort and cooperation of all members of the Panel. The author of the sections on history, abundance, properties, sources and distribution, and sampling and analysis of chromium was Dr. James O. Pierce, II. The epidemiologic aspects were covered by Dr. Philip E. Enterline. The dermatologic problems were discussed by Dr. Donald J. Birmingham. Dr. Walter Mertz drafted the material on biologic and metabolic interactions of chromium and on chromium in nutrition. The material on clinical studies was provided by Dr. P. Lesley Bidstrup. Dr. Baetjer wrote the sections on experimental exposure of animals, on current standards for industrial exposures and for ambient air and water pollution, and, with the cooperation of Dr. Mertz, on the effects on vegetation. The section on chromium in aquatic species was taken directly from the National Academy of Sciences report, *Water Quality Criteria* (in press), and a 1971 Naval Research Laboratory review, *Hexavalent Chromium: Toxicological Effects and Means for Removal from Aqueous Solution.*

The preparation of the document was assisted by the comments of the Committee on Biologic Effects of Atmospheric Pollutants* and, in particular, Dr. W. Clark Cooper, who served as Associate Editor, and Dr. David M. Anderson, who served as Committee liaison. The Panel is indebted to the anonymous reviewers, who offered many useful comments on the original typescript.

Dr. Robert J. M. Horton of the Environmental Protection Agency (EPA) was of invaluable assistance in obtaining documents and translations and in providing counsel. Dr. Douglas I. Hammer was EPA liaison officer. Informational assistance was provided by the National Research Council (NRC) Advisory Center on Toxicology, the National Academy of Sciences Library, the National Library of Medicine, the National Agricultural Library, the Library of Congress, and the Air Pollution Technical Information Center.

The Panel is greatly indebted to Mr. John Redmond, Jr., the responsible staff officer in NRC, for his invaluable assistance and splendid cooperation throughout this project. Mr. Norman Grossblatt, Editor of the Division of Medical Sciences, edited the manuscript.

Acknowledgment is made of the assistance given by the Environmental Studies Board, NAS–NAE, and divisions of the National Research Council.

*The Committee's title has since been changed to Committee on Medical and Biologic Effects of Environmental Pollutants; its purpose and scope remain the same.

Contents

Introduction

As civilization has progressed, man has used ever greater quantities of chemicals, has learned to prepare new compounds not found in nature, and has devised many different chemical processes. These developments have led to great concern over the public-health consequences of accidental release or waste disposal of these possibly harmful chemicals into the environment. The problem has been intensified by the development of more sensitive methods of detection and analysis.

Because some chromium compounds have been found to produce harmful effects in man through industrial exposures, it has become important to determine the concentrations, forms, and health effects of chromium in the ambient air.

This document is the result of a critical evaluation of the literature available to the Panel on Chromium up to July 1, 1972, on the biologic and health effects of chromium. As will become apparent in the following pages, the literature does not provide conclusive answers to many of the important questions on the biologic effects of chromium. The final chapter lists categories of future research that the Panel feels warrant high priority.

1

Properties of Chromium

The nuclear, physical, mechanical, and chemical properties of chromium are very complicated and require a more detailed discussion than is appropriate for this report. This chapter briefly summarizes the more important properties of chromium.

HISTORY

Chromium, unlike many metals, was unknown to ancient peoples. There are no records of the use of chromium or its compounds, possibly because of the unattractiveness, refractoriness, or scarcity of the various minerals containing it.[338]

Chromium was discovered, by Nicolas-Louis Vauquelin in 1797, in Siberian red lead ore (crocoite), $PbCrO_4$. In 1798, the new metal was isolated by reduction of chromium oxide, CrO_3, with charcoal at high temperature. In the same year, the green color of a Peruvian emerald was found to be due to the presence of chromium. Fourcroy and Hauy suggested the name "chrome" for the element (from the Greek *chroma*, color) because of its many colored compounds.[274]

A few years after the discovery of chromium, the commercial process of manufacturing chromates (roasting chromite with lime and soda ash) was developed. Andreas Kurtz, a pupil of Vauquelin, started manufacturing chromium chemicals in London in 1816. In 1822, the operations were moved to Manchester, where bichromates and chromium pigments, such as lead chromate, were produced. Mordant dyeing with chromium compounds was begun in 1820. Chrome tanning was invented by Friedrich Knopp in 1858 and was commercialized in 1884, after the patenting of the chrome-tanning process of Augustus Schultz. The use of chromite, $FeOCr_2O_3$, as a refractory started in 1879; its use in metallurgy started to become important around 1910. Chrome plating, in the modern form, dates from about 1926.[274]

In the United States, chromium became an economically important element in 1827, when relatively large deposits of ore were discovered near Baltimore, Maryland. During the years 1827–1860, the Baltimore region produced practically all the world's supply of chromite; after the discovery, in 1848, of large deposits of chromite near Brusa, Asia Minor, this monopoly no longer existed.

The increased usage of chromium, due primarily to technologic advances within the last two decades, has made this element of considerable importance.

Chromium is now recognized as an essential trace element in man. The first conclusive evidence was obtained by Mertz and Schwarz in 1955.[210,217] The biologic importance of chromium is discussed in detail in Chapter 5.

ABUNDANCE

Chromium is fairly abundant in the earth's crust, ranking fourth among the 29 elements of biologic importance and seventeenth among all the nongaseous elements. It is more abundant than cobalt, copper, zinc, molybdenum, lead, nickel, and cadmium.[292] Chondrites, the most common of meteorites, contain chromium at about 3,000 μg/g (ppm).[2]

In seawater, chromium occurs in rather low concentrations, ranking twenty-seventh or lower among the elements. It is thought that little chromium remains in solution in seawater.[294] The only important chromium ore is chromite, a mineral of the spinel group, with the formula $(Fe, Mg)O(Cr, Al, Fe)_2O_3$. The theoretical end member $FeOCr_2O_3$ would contain 68% chromic oxide, Cr_2O_3, and 32% ferrous oxide, FeO. The highest grades of ore contain 52–56% chromic oxide and 10–26% ferrous oxide, with varying amounts of other substances, including magnesia, alumina, and silica.[274]

The chief producers of chromite have been the U.S.S.R., Union of

South Africa, the Philippines, Turkey, and Southern Rhodesia. Chromite deposits are also found in many other countries, including the United States, Albania, Cuba, Brazil, Japan, India, New Caledonia, Pakistan, Iran, and the Malagasy Republic.[15] Although chromite deposits do exist within the continental United States, the concentrations have been such that it is currently not economically feasible to mine and reclaim the chrome ore. Thus, the United States is not a producer of chrome ore.

Crocoite has been found in small quantities in the U.S.S.R., Brazil, Hungary, the Philippines, and Tasmania. Chromium occurs in minor amounts in many minerals, in which (as Cr^{+3}) it replaces Fe^{+3} or Al^{+3}—e.g., in the chromium tourmalines, chromium garnets, chromium micas, and chromium chlorites.[265] The true emerald is a form of beryl (beryllium aluminum silicate), colored green by incorporation of a small amount of chromium in place of aluminum.[360] The color of the ruby is also due to a trace of chromic oxide, which distinguishes the gem from the common crystalline corundum (alumina).

NUCLEAR PROPERTIES

Chromium, with an atomic number of 24 and an atomic weight of 51.996, is a mixture of four stable isotopes of mass numbers 50 (4.31%), 52 (83.76%), 53 (9.55%), and 54 (2.38%). The thermal-neutron capture cross sections (O_c) are 17, 0.8, 18, and 0.38, respectively.[183]

Five radioisotopes are known. Two others (mass numbers 46 and 47) have been reported, but their existence is uncertain. The radioisotope commonly used in tracer work is chromium-51.

PHYSICAL AND MECHANICAL PROPERTIES

There are sufficient data in the literature pertaining to the physical and mechanical properties of chromium.[10,14,164,288,315,318] These data—such as entropy, ground states, and orbital radii—were compiled in a long table for inclusion in this report, but excluded on the grounds that such data would be of little value to the reader. Estimates of numerical values of many of the physical properties of chromium—such as melting point, boiling point, and heat of sublimation—vary, owing primarily to the metal's refractoriness and its reactions with many other substances at high temperatures. Furthermore, many of its properties vary with the purity of the chromium sample, which is often unknown or at least unstated. The mechanical properties depend so heavily on impurity content, test conditions, and other factors (including the brittleness of the metal itself) that the values attached to such properties are of questionable use.[274]

CHEMICAL PROPERTIES

Chromium is a member of group VIB of the periodic table. It has oxidation states ranging from Cr^{-2} to Cr^{+6}, but it most commonly occurs as Cr^{0}, Cr^{+2}, Cr^{+3}, and Cr^{+6}. Divalent chromium, however, is relatively unstable, being rapidly oxidized to the trivalent form; thus, only two forms—trivalent and hexavalent—are found in nature. The oxidation potential of hexavalent to trivalent chromium is strong, and it is highly unlikely that oxidation of the trivalent form could occur *in vivo.*[297] The hexavalent form of chromium, almost always linked to oxygen, is a strong oxidizing agent.

The only important chromium ions are chromates and dichromates, which are easily reduced to trivalent chromium in acid solution and in the presence of organic matter.[292] Commercially available forms of chromium include Cr^{+3} in dilute acid (usually hydrochloric acid) and chromate.[274]

The trivalent state is the most stable and important oxidation state of chromium. In this state, it has a strong tendency to form complexes whose ligand rates of exchange are low (half-times of several hours). Trivalent chromium forms octahedral complexes of coordination number 6, and a large number of complexes are known—water, ammonia, urea, ethylenediamine, halides, sulfate, and organic acids. It is largely because of this kinetic inertness that these compounds persist for relatively long periods in solution, even in conditions in which they are thermodynamically very unstable. Chromic ion does not exist in solution. It forms complexes with water and other anions in an acid solution. In an alkaline solution, it olates by forming polynuclear compounds, which precipitate in time. Olation is enhanced by alkali and heat to 120 C.[292]

The magnetic properties of the octahedral trivalent chromium complexes are uncomplicated. From a simple orbital splitting diagram, it follows that all such complexes must have three unpaired electrons, irrespective of the strength of the ligand field. This has been confirmed in all known mononuclear complexes. More sophisticated theory further predicts that the magnetic moments should be very close to but slightly below the spin-only value of 3.88 BM; this, too, is observed experimentally.[68 (pp. 818-833)]

The spectra of trivalent chromium complexes are well understood in their main features. Three spin-allowed transitions are expected, and these have been observed in a considerable number of complexes.[68 (pp. 818-833)]

All hexavalent chromium compounds except chromium hexafluoride, CrF_6, are oxo-compounds. Acid solutions of dichromate are powerful oxidizing agents; e.g.,

$$Cr_2O_7^{-2} + 14H^+ + 6e^- = 2Cr^{+3} + 7H_2O, E^0 = -0.13\ V.$$

Hexavalent chromium does not give rise to the extensive and complex series of polyacids and anions characteristic of the somewhat less acidic oxides. The reason for this is perhaps the greater extent of multiple bonding for the smaller chromium ion. Other than the chromate and dichromate ions, there are no oxyacids or anions of importance, although trichromates, $Cr_3O_{10}^{-2}$, and tetrachromates, $Cr_4O_{13}^{-2}$, have been reported.[68 (pp. 818-833)]

Another type of trivalent chromium compound of note is exemplified by chromyl chloride, CrO_2Cl_2, a deep red liquid (boiling point, 117 C) formed by the action of hydrogen chloride on hexavalent chromium oxide,

$$CrO_3 + 2HCl \rightarrow CrO_2Cl_2 + H_2O,$$

by warming dichromate with alkali metal chlorides in concentrated sulfuric acid,

$$K_2Cr_2O_7 + 4KCl + 3H_2SO_4 = 2\ CrO_2Cl_2 + 3K_2SO_4 + 3H_2O,$$

and in other ways. It is photosensitive and vigorously oxidizes organic matter, but is otherwise rather stable. It is hydrolyzed by water to chromate ion and hydrochloric acid.[68 (pp. 818-833)]

The hexavalent form of chromium appears to be relatively stable in water, probably because of the low concentrations of reducing materials.[73,109] The trivalent form is associated mainly with particulate matter, which suggests that organic particles may reduce and bind the element, leaving the hexavalent form in solution.[71] Other data suggest that chromium may be absorbed on clay particles.

Chromium metal resists attack by a wide variety of chemicals at normal temperatures, but reacts with many of them at high temperatures.[274] It reacts with several of the common acids with evolution of hydrogen, and it reacts slowly with dilute sulfuric acid.[10] It dissolves in aqueous hydrogen fluoride, hydrogen chloride, hydrogen bromide, and hydrogen iodide, and forms Cr^{+2} (in the absence of air); if pure chromium is used, very little Cr^{+3} is formed.[172,196] It is not attacked by phosphoric acid, and it resists attack by many organic acids, including formic, citric, and tartaric; but it is attacked slowly by acetic acid.[10,288]

Chromium is insoluble in nitric acid, fuming nitric acid, and aqua regia.[10,288] It is made passive and rendered relatively nonreactive by these acids and by other oxidizing agents, such as chlorine, bromine, and solutions of hydrochloric acid and chromium oxide. It is also slowly passivated by superficial oxidation in air, although this is far less effective in causing passivation than the other agents mentioned. Passive chromium can be made active by a reduction process, such as treatment with hydrogen gas, or by immersion in dilute sulfuric acid and touching with zinc below the surface of the acid.[10]

Although passive chromium acts somewhat like a noble metal and is not attacked by mineral acids, it is very active in the nonpassive state. In aqueous solution, it can displace copper, tin, and nickel ions from their salts. Passivation has been attributed to surface absorption of oxygen or formation of an oxide coat, but it must be concluded that no adequate explanation has been proposed.

2

Distribution of Chromium in the Environment

Adequate discussion of the distribution of chromium in the environment rests on the assumption that the data in the published literature are reliable. In the case of chromium and other metals, this assumption cannot be made—primarily because the analytic methods and sampling techniques used by investigators in the past have been, as a whole, unreliable and highly variable. The discrepancies in analytic data are particularly evident in literature discussions of chromium concentrations in the so-called trace range—i.e., in parts per billion. This is especially true of data on chromium concentrations in water, air, and biologic materials. The geochemical data—such as those on rocks and soil—may be somewhat more reliable. For these reasons, it is difficult to assess properly the chromium concentrations reported in this chapter. (These statements should not be construed as reflecting on the investigators whose data are used here. In reviewing the data, one should keep in mind that the best analytic methods available were used; but we now know that those methods were not adequate, compared with recent developments and refinements in analytic methodology. This subject is discussed in greater detail in the appendix.)

Chromium is an essential trace element. Since sensitive methods were developed, chromium has been found in almost all living things—vegetation, animals, and people. Chromium is also found in the air, soil, and water. Adding chromium to soils that are deficient in chromium can increase plant growth, but large amounts are poisonous. Reported chromium concentrations vary from author to author. Many older papers reported an absence of chromium in biosystems; but, as detection sensitivity increased, the presence of chromium in more and more materials was reported. When work is done near a method's limit of detection, erroneous results are common (owing to interferences and equipment problems that are not important at higher concentrations). Sample treatment and contamination are also more critical in dealing with very low concentrations.

The following sections contain data from the published literature. The numbers used must be viewed in the light of recently discovered methods of measurement of chromium, which may lead to justified suspicion of the accuracy and precision of older data.

Elemental chromium is not found in nature. The only important commercial chromium mineral is chromite, $FeOCr_2O_3$, which is seldom found in pure form. The FeO in chromite is often replaced by MgO, resulting in $MgOCr_2O_3$. Table 2-1 shows typical concentrations of chromium in rocks and other materials.

EARTH'S CONTINENTAL CRUST

The range of chromium concentration in the continental crust is 80–200 ppm, with an average of 125 ppm; however, concentrations as high as 370 ppm have been reported.

SOIL

Soils and rocks contain small, but varying, amounts of chromium, usually as chromic oxide. The concentration of chromium in soil has been reported to vary from a trace to 5.23% (Puerto Rico).[268] Studies by Shacklette and others on the concentration of chromium in soils throughout the United States led to a geometric mean of 37 ppm, on the basis of 863 soil samples.[304] Concentrations of chromium in soil are rarely high, usually being between a trace and 250 ppm, as chromic oxide.[267] Chromium is said to be particularly concentrated in soils derived from basalt or serpentine.[42]

Higher concentrations of chromium are usually found in ultramafic igneous rocks, in shales and clays, and in phosphorites. The chromium in phosphorites used as fertilizers is an important source of man-induced contamination of soils with chromium. The concentration of

TABLE 2-1 Abundances of Chromium in Various Materials[a]

	Chromium Content, ppm	
Material	Range Usually Given	Average
Continental crust	80–200	125
Ultramafic igneous rocks	1,000–3,400	1,800
Basaltic igneous rocks	40–600	220
Granitic igneous rocks	2–90	20
Shales and clays	30–590	120
Limestones	–	20
Sandstones	–	35
Soils[b]	10–150	40
Phosphorites	30–3,000	300
Coals	10–1,000	20[b]
Petroleum	–	0.3[b]
Seawater	0.00004–0.0005	–
Marine plants	–	1
Marine animals	0.2–1	–
Land plants	–	0.2
Insects	trace	–
Mammals	–	0.3
Fresh water	0.0001–0.08	0.00018

[a]Derived from Bowen.[42]
[b]Data from Bertine and Goldberg,[26] who report chromium at 10 ppm for coal and 0.3 ppm for petroleum. Other sources have reported chromium concentrations in coal as high as 60 ppm. Chromium content of soil depends heavily on type of parent rock and degree of contamination by man.

chromium in phosphorites ranges from 30 to 3,000 ppm, with an average of 300 ppm (see Table 2-1). The phosphorites from Idaho, Wyoming, and Utah contain chromium at an average of about 1,000 ppm.[191]

High chromium content has been associated with infertility of some soils. For example, the soil in a small area near Rockville, Maryland, which supported minimal vegetation, contained chromium at 1,000–3,900 ppm, as chromic oxide.[268]

Some soils, especially those formed from calcareous rocks, appear to be deficient in available chromium. Deficiency in soils must be defined not as a concentration of total chromium below some arbitrary limit, but as a condition of the soil that results in insufficient chromium uptake by the plants growing on it.[210]

WATER

The chromium content of river waters has been carefully studied with spectrographic techniques by the United States Geological Survey.[127]

A survey of chromium in 15 rivers of North America reported concentrations ranging from less than 0.7 ppb (Sacramento River) to 84 ppb (Mississippi River). Most of the samples yielded between 1 and 10 ppb. Chromium is significantly more concentrated in Atlantic coastal river waters and significantly less so in Gulf and Pacific river waters than in the average.[87,321] This probably reflects the sizes of the different watersheds and the compositions of soils drained by them.

Chromium is generally less concentrated in seawater than in rivers and wells, with concentrations well below 1 ppb. Measured concentrations have been 0.04–0.07 ppb in samples taken near the coast of Japan,[159] 0.13–0.25 ppb near England,[192] 0.2–0.4 ppb in the Ligurian Sea,[109] and 0.46 ppb in the Irish Sea.[60] Both trivalent and hexavalent forms have been shown to exist in seawater. In one investigation, an increased proportion of chromium in the trivalent form was found in deeper waters.[109] It appears likely that the hexavalent chromium on the ocean surface is constantly being diminished by interaction with organic particles and slowly settles, either because of adsorption on particulate matter[71] or because of the formation of insoluble hydroxides.[174] The content of hexavalent chromium is replenished by the considerable amounts of chromates delivered to the oceans by the rivers. It has been estimated that 6.7×10^6 kg of chromium are added to the oceans each year.[42] Thus, much of the chromium lost from the land by erosion and mining is eventually deposited on the ocean floor.[210]

Municipal drinking water in the United States, according to a 1962 study, may contain no detectable chromium or up to 35 ppb, with a median of 0.43 ppb.[86] The public water supplies probably do not reflect accurately the chromium content of the waters from which they are drawn. There may be wide variations in chromium content between the products of different waterworks of one city. For Washington, D.C.,

TABLE 2-2 Chromium in Drinking Water and Surface Waters in United States[a]

No. and Type of Samples	Chromium Concentration, ppb		Comment
	Minimum	Maximum	
700 surface	*b*	50	11 samples in range of 6–50 ppb
2,595 drinking	*b*	79	5 samples higher than 50 ppb
1,577 drinking	1	112	mean, 9.7 ppb
380 drinking	1	29	mean, 7.5 ppb

[a]Data from Durum *et al.*,[88] Kopp,[171] and U.S. Department of Health, Education, and Welfare.[346]
[b]Not reported.

for example, the two supplies contain 0.49 and 6.6 ppb—a difference probably introduced by processing. The water of Milwaukee, Wisconsin, was recently rechecked because of its high reported chromium content of 35 ppb.[86] The chromium concentration measured by atomic-absorption spectroscopy was less than 0.8 ppb in 1966, after the waterworks operating in 1962 had been closed.[210]

Table 2-2 summarizes several studies of the chromium content of drinking water and surface waters in the United States.

FOOD

In long-term balance studies, Tipton[332] has analyzed diets and excreta for 17 elements. His subjects were consuming three times as much chromium as was present in a hospital diet (Table 2-3) and much more than normal. Variations in average daily intake, however, are wide, from 5 μg well into the hundreds of micrograms[292] (see Table 2-3).

Foods vary considerably in chromium content (Table 2-4). The largest sources are meats, mollusks and crustaceans, vegetables, and unrefined sugar. In general, fish, vegetable oils, and fruit have been reported to contain smaller amounts of chromium. Schroeder has estimated the relative contributions of food, water, and air to average daily intake (Table 2-5).[292] Kirchgessner and his group[165,166,365,366] report chromium concentrations ranging from 80 ppb (in potatoes) to 590 ppb (in hay). The chromium content of hay was found to increase sharply at the time of ripening of the seeds. Vegetable products for human consumption contain much less chromium (20–50 ppb)—for example, fruits, 20 ppb, and grains and cereals (excluding fats), 40 ppb.[210]

TABLE 2-3 Daily Intake of Chromium in Various Diets

Diet or Food (quantity)	Chromium, μg	Area	Reference
Institutional	78	Vermont	Schroeder *et al.*[300]
Institutional	52	Syracuse, N.Y.	Levine *et al.*[189]
Hospital	101	Vermont	Schroeder *et al.*[300]
Ad libitum	65 (5–115)	Syracuse, N.Y.	Levine *et al.*[189]
Dinner (200 g)	64	Rome, Italy	Mertz[210]
Self-selected	130–140	Japan	Murakami *et al.*[226]
Self-selected	200–400	Cincinnati, Ohio	Tipton *et al.*[333,334]
Vegetarian	11– 55	India	Joseph *et al.*[162]
Sugarcane (500 g)	100	S. Africa	Mertz[210]
Lentil soup (200 ml)	38	Egypt	Carter *et al.*[55]
Maize (1 liter)	10	Nigeria	Hopkins *et al.*[146]
Milk powder (1 liter)	18	Jordan	Hopkins *et al.*[146]

TABLE 2-4 Chromium Concentrations of Various Foods[a]

Food	No. Samples	Chromium Concentration, μg/g (wet wt)	
		Mean	Range
Condiments	6	2.71	0.01–10
Meat	9	0.13	0.03– 0.27
Fish	2	0.02	0.01– 0.02
Seafood	9	0.15	0 - 0.44
Vegetables	36	0.18	0 - 3.62
Grains	15	0.13	0 - 0.52
Cereal products	10	0.22	0.05– 0.23
Vegetable oils	12	0.08	0.03– 0.23
Fruits	9	0.01	0 - 0.2
Sugars and syrups	65	0.10	0 - 1.1
Frozen TV dinners	3	0.09	0.05– 0.13

[a]Data from Schroeder,[292] Schroeder *et al.*,[300] and Sullivan.[327]

NONFOOD VEGETATION

Chromium has been found in many plant tissues (Table 2-6).[42] More chromium is taken up by plants growing on soils formed from ultrabasic rocks than by those on silica or calcareous rocks, by a factor of 5–40.[297] Furthermore, higher plants—conifers, deciduous trees, and shrubs—take up less chromium than do lower plants—lichens, mosses, ferns, and grasses. The lower plants have mean chromium concentrations of 100–650 ppm (ash), and the higher plants, 18–73 ppm (ash) when growing on ultrabasic soils (generally serpentine or talc) that

TABLE 2-5 Estimated Mean Daily Intake of Chromium in Man[a]

Source	Chromium Intake, μg	
	Hospital Diet	Self-Selected Diet
Food	100 (5–500)	280
Water	1 (0–84)	4
Air	0.04 (0–0.8)	0.28

[a]Data from Schroeder.[292] Estimates of the contributions of chromium from food, water, and air vary significantly from author to author. Those in this table are only examples. Possibly the best tentative conclusion that can be drawn from such data is that foods vary considerably in chromium content and probably constitute the greatest source of intake in man.

TABLE 2-6 Chromium in Plants[a]

Plant Type	Chromium Concentration, ppm (dry wt)
Marine plants	1
Land plants	0.23
Plankton	3.5
Brown algae	1.3
Bryophytes	2
Ferns	0.8
Gymnosperms	0.23
Fungi	1.5

[a]Derived from Bowen.[42]

contain chromium at 2,000–6,000 ppm. Lichens and mosses take up chromium at 39–48 ppm from silica soils that contain 140 ppm (range, 10–330 ppm), whereas other plants take up only 4.9–7.6 ppm from the same soils.

Hanna and Grant found chromium in 38 of 43 samples of 13 trees and shrubs in New Jersey—usually 0.2–0.6 ppm (dry wt). Most other investigators have found similar concentrations. Analysis of wild vegetation in Vermont yielded an average concentration of 0.39 ppm (dry wt).[134]

Vegetables of 25 botanic families were found to contain 10–1,000 μg of chromium per kilogram of dry matter, with most plants falling between 100 and 500 μg/kg.[327]

AIR

Current data for ambient chromium concentrations in the air of major urban areas are apparently inadequate to depict trends accurately. Cursory examination of these data for 20 of the larger (or more industrialized) cities reveals what appears to be, on the whole, a slight downward trend. But such trends will not be reliable until the presently contemplated control of particulate emissions has been implemented.

Table 2-7 covers the period 1960–1969, beginning at the point when analytic techniques had become somewhat standardized and large numbers of samples were being analyzed, and extending into the era of automation, by which time samples can be routinely analyzed by rather precise methods. The table shows only the cities that consistently had high annual mean concentrations. Most of the remaining 200 or so urban stations will normally have annual mean concentra-

TABLE 2-7 National Air Surveillance Networks Chromium Data[a]

	Annual Mean Concentration of Chromium in Air, $\mu g/m^3$									
City	1960	1961	1962	1963	1964	1965	1966	1967	1968	1969
Los Angeles	0.030	0.018	0.026	0.015	0.0	0.0	0.03	0.0	0.0	0.02
Denver	0.0	0.0	0.009	0.007	0.005	0.0	0.03	0.0	0.0	–
District of Columbia	0.0	–	0.016	0.006	0.016	0.0	0.0	0.0	0.0	–
Atlanta	0.03	–	–	0.007	0.0	0.002	0.0	0.0	0.021	0.0
Chicago	–	0.039	0.0	0.013	0.014	0.0	0.04	0.0	0.023	0.02
East Chicago	–	–	–	–	0.033	–	–	–	0.037	0.06
Indianapolis	–	0.042	0.016	–	0.008	–	–	–	0.017	0.01
Baltimore	0.301	0.108	0.051	–	0.069	0.018	–	–	0.051	0.10
Detroit	0.05	0.03	0.016	0.010	0.014	0.0	0.05	0.025	0.043	0.02
St. Louis	0.0	0.0	0.021	0.020	0.007	0.0	0.04	0.0	–	0.02
Newark	0.03	0.046	0.020	0.0	0.020	0.0	0.03	0.0	0.0	0.02
New York	0.04	0.053	0.027	0.016	0.006	0.0	0.0	–	–	0.02
Cleveland	0.05	0.04	0.017	0.020	0.014	0.0	0.03	0.0	0.0	0.02
Youngstown	0.051	–	0.008	0.012	–	–	–	–	0.045	0.03
Philadelphia	0.049	0.024	0.018	0.013	0.011	0.0	0.05	0.0	0.0	0.02
Pittsburgh	0.197	0.07	0.041	0.023	0.021	–	0.018	0.028	0.05	0.05
Reading	–	–	–	–	–	–	–	–	0.063	0.06
Chattanooga	–	0.041	–	–	0.015	–	–	–	0.017	0.02
Houston	0.03	0.0	0.0	0.003	0.004	0.006	0.0	0.0	0.0	–
Charleston	0.05	0.132	0.0	0.0	0.044	0.04	0.06	0.03	0.0	0.04

[a]Data from U.S. Department of Health, Education, and Welfare[340,343-345] and U.S. Environmental Protection Agency.[348,349]

tions of 0.01–0.03 μg/m^3, although they will occasionally have concentrations below the detectable minimum. Several will seldom have samples containing more than the detectable minimum (0.01 μg/m^3).

Annual mean chromium content at most rural stations seldom reached 0.01 μg/m^3, except in a few instances, such as the Puerto Rico station in 1959 (0.037 μg/m^3) and the Florida station in 1963 (0.01 μg/m^3). In these areas, the air sometimes appears to take on some of the aspects of urbanization.

COMBUSTION

Table 2-7 summarizes reported concentrations of chromium. Particles emitted from coal-fired power plants contained 2.3–31 ppm, depending on the type of boiler firing, and the emitted gases contained 0.22–2.2 mg/m^3; these concentrations were reduced by fly-ash collection to 0.19–6.6 ppm and 0.018–0.5 mg/m^3, respectively.[327]

Table 2-8 shows the results of one study of chromium emission from coal-fired power plants.

Wood contains chromium, and it is likely that the burning of wood in fireplaces and campfires may contribute a small amount of chromium to the air. Forest fires would be a large source. Leaf burning and incineration of rubbish undoubtedly put a small amount of chromium into the atmosphere, inasmuch as leaves contain measurable concentrations.[292]

TABLE 2-8 Typical Chromium Emission From Coal-Fired Power Plants[a]

		Chromium Emission, mg/m^3	
Type of Boiler Firing	Ash in Coal (as fired), %	Before Fly-Ash Collection	After Fly-Ash Collection
Vertical	20.2	0.22	0.02
Corner	14.9	1.90	0.13
Front-wall	10.3	1.10	0.16
Spreader-stoker	8.4	0.45	0.35
Cyclone	7.7	1.88	0.50
Horizontally opposed	8.2	2.22	0.41

[a]Derived from Cuffe and Gerstle.[69] These data are illustrative only. Actual results vary according to chromium concentration of coal being burned.

CYCLING OF CHROMIUM IN THE BIOSPHERE

Almost no data are available on the total ecologic cycling of chromium in the environment. Much additional study needs to be undertaken in this crucial area.

The trivalent and zero-valent chromium in air should not undergo any reaction. The hexavalent chromium in air could eventually react with dust particles and other pollutants to form trivalent chromium, but the reactions and forms of chromium in air have not been extensively studied.

Chromate may enter municipal sewage in many different ways. It occurs perhaps most commonly in plating wastes, although it may have its source in tanning operations, in waters given chromate corrosion-inhibition treatment, or in aluminum-anodizing wastes. Because chromate retained in a sewage-treatment plant is reduced to chromic chromium, the effect of this form of chromium on sewage treatment is pertinent.[341]

In a study performed at the Bryan, Ohio, sewage-treatment plant, a slug of approximately 21 kg of chromium was studied. It was found to have little effect in plant operations; approximately 80% was removed from the effluent stream. The concentration of chromium in the stream that received the effluent was observed to increase from less than 0.1 mg/liter to 0.3 mg/liter.[341]

The chromium content of sewage sludges is of particular importance, especially in view of the widespread advocacy of their use for improving soils. Table 2-9 lists concentrations of chromium found in sewage sludges in various locations.

TABLE 2-9 Reported Concentrations of Chromium in Sewage Sludges

		Chromium Content, ppm			
Location	No. Samples	Range	Average	Median	Reference
England and Wales	42	40–8,800	980	250	25
Lexington, Ky.	6	2,100–3,200	–	–	264
Kentucky	12	200–9,100	3,400	3,100	264
Chicago	2	360–1,680	–	–	330

3

Industrial Uses of Chromium

The pollution of air by chromium and its compounds comes primarily from industrial use and product use. There has been no mining of chromite ore in the United States since 1961, because higher-grade ore can be purchased from foreign countries at a lower cost. Approximately 57% of the imported chromite ore is used in the metallurgic industry, 30% in refractory materials, and 13% in the chemical industry.[327]

The metallurgic industry uses ores containing at least 50% chromic oxide for ferrochrome, which is used primarily in stainless and alloy steels. The chromium used in the metallurgic industry is mainly trivalent, or in the zero state. Metallurgic-grade chromite ore is usually converted into one of several types of ferrochromium or chromium metal that are alloyed with iron or other elements (usually nickel and cobalt). From these alloys are produced a great variety of useful steels. Over 60% of the chromium used in the metallurgic industry is used in making stainless steel; the remainder is used in austenite steels, high-speed steels, other alloy steels, high-temperature steels, and nonferrous alloys. Current ferrochrome manufacturing practice uses 75% of the highest-quality ore (with a chromium : iron ratio in excess of 3 : 1 being desired), which is available mostly from the U.S.S.R., Turkey, and Rhodesia. Stockpile releases have been supplementing requirements.

The refractory industry uses ores containing approximately 34%

chromic oxide and high alumina content for melting-furnace linings, because chromite has a high melting point (2040 C) and is chemically inert. In addition to their use as chromite bricks or magnesia–chrome bricks, chrome refractory materials may be used as coatings to close pores and for joining bricks within the furnace.[292] This use of chromite is important but declining, as open hearths are replaced by basic-oxygen furnaces. The chromium used in the refractory industry is trivalent as chromite ore.[292]

The chemical industry uses ore containing approximately 45% chromic oxide for preparation of sodium chromate and sodium dichromate, from which most other chromium chemicals are produced. Chromium chemicals are used as tanning agents, pigments, catalysts, and plating and wood preservatives.[327] This ore is primarily from South Africa and is plentiful.[229]

For current and projected trends in the use of chromium, see Tables 3-1 through 3-4.

TABLE 3-1 United States Consumption of Chromite by Industry of Usage[a]

	Chromite Consumption, 1,000 tons						Metallurgic Consumption, % of Total Chromite
	Metallurgic		Refractory		Chemical		
Year	Total[b]	Chromium Content	Total[c]	Chromium Content	Total	Chromium Content	
1952	677	218	340	80	147	45	58
1953	743	236	441	–	152	46	56
1954	502	159	278	65	133	41	55
1955	994	316	431	101	159	49	63
1956	1,212	388	475	112	160	50	66
1957	1,177	379	435	104	148	46	67
1958	778	250	312	75	131	41	64
1959	796	252	379	91	162	50	60
1960	665	211	391	93	164	47	54
1961	662	211	375	89	163	51	55
1962	590	188	365	87	176	55	52
1963	632	210	368	87	187	58	53
1964	832	279	430	99	189	58	57
1965	907	309	460	109	217	67	57
1966	828	281	439	104	194	60	57
1967	866	294	310	72	179	55	64
1968	804	273	311	72	202	62	61

[a]Derived from National Research Council.[229]
[b]Some part of the total, usually between 10,000 and 20,000 tons, was added directly to steel; the balance was used to make ferroalloys and chromium metal.
[c]A small quantity, usually between 5,000 and 10,000 tons, was in direct furnace repairs; the balance was used in making brick and other refractory products.

TABLE 3-2 Forecast Growth in Chromite and Chromium Consumption in the United States[a]

Use	Chromite Consumption After Allowance for Scrap, 1,000 tons			Chromium Content After Allowance for Scrap, 1,000 tons	
	1968	1973	% Change	1968	1973
Stainless steel	525	659	+ 26	163	205
Alloy steel	125	157	+ 26	39	49
Tool steels[b] (all types)	16	19	+ 19	5	6
High-temperature and nonferrous alloys	61	87	+ 43	19	27
Foundries, metallurgic	61	84	+ 38	19	26
Miscellaneous metallurgic applications[c]	6	10	+ 67	2	3
SUBTOTAL (metallurgic)	794[e]	1,016	+ 28	247	316
Foundries, facing-sand	26	65	+150	8	20
Refractories	310	250	- 19	74	60
Chemicals[d]	226	254	+ 12	70	78
TOTAL	1,356	1,585	+ 17	399	474

[a]Derived from National Research Council.[229]
[b]Based on production of 96,000 tons of tool steel with an average chromium content of 6%.
[c]Includes cutting and wear-resistant materials, welding and hard-facing rods, and use in other steels.
[d]Consumption in chemicals market in 1968 was estimated at 149,000 tons of sodium dichromate equivalent. One ton of $Na_2Cr_2 \cdot 2H_2O$ requires 1.4 tons of ore, based on 80–85% recovery.
[e]Calculated as 50% ore, but small quantities of chemical-grade ore (44–45% chromium) and refractory-grade ore (34–37% chromium) are used.
Notes:
1. The projection includes allowance for losses during use of the ferroalloys in metallurgic processing.
2. The projection includes an additional 10% loss for processing chromite into ferroalloys.
3. Average assay of ore for metallurgic uses is 50% Cr_2O_3.
4. Average assay of ore for refractory use is 35% Cr_2O_3, and no processing loss is assumed.
5. Average assay of ore for chemicals and facing-sand uses is 45% Cr_2O_3.

A majority of the chromic acid produced is used for chrome plating. This process results in chromic acid aerosols released by bubbles of hydrogen in the solution, which are exhausted in the air. It has been found that industrial workers dealing with the following were exposed to hexavalent chromium: tanning, primer paints, pigments, graphic arts, printing and reproducing, fungicides, wood preservatives, and corrosion inhibitors.[292]

TABLE 3-3 Chromium Usage Trends by Major Product[a]

Use	Estimated Chromium Usage in 1968, tons	Usage Trend, 1968–1973	Potential Substitutions
Stainless steel	263,000	Increasing	No major substitutes obvious for chemical process equipment of high-temperature applications requiring corrosion or oxidation resistance; in small quantities (5% of total stainless capacity), copper–nickel or titanium-base alloys could be substituted at higher cost
Alloy steel	46,000	Increasing	Main markets in construction and automotive industries; substitutions usually feasible
Refractories	74,000	Decreasing	Decrease due to rapid decline in use of open-hearth furnace for steel manufacture; magnesite can be substituted in some applications
Chemicals	70,000	Increasing	Some uses (including pigments, plating, metal treatment, and catalysis) will increase; substitution in major uses usually feasible at cost or performance penalty
Foundry applications; iron and steel castings	31,000	Increasing	Production of steel castings and increasing use of chromite as facing sand responsible for most of increase; zircon sand could be substituted at higher cost

[a]Derived from National Research Council.[229]

TABLE 3-4 Chromium Usage Trends by Industry[a]

Market	Estimated Chromium Usage in 1968, tons	Usage Trend, 1968–1973	Comments
Motor vehicles	89,000	Increasing	All applications for automotive use appear to be rising
Aircraft	22,000	Decreasing	Alloy and superalloy consumption has been, and continues to be, on the decline
Marine transportation	1,000	Decreasing	Alloy use is dropping irregularly
Appliances, utensils, service machinery	19,000	Increasing	Population growth alone means increase
Clothing (leather)	6,000	Decreasing	Synthetics and hide shortages plus increased imports mean a drop in usage of chrome tanning chemicals
Electric and electronics	10,000	Increasing	Increasing general importance of this industry presages growth in chromium usage
Process industry	13,000	Increasing	Use should continue to increase with population growth
Heavy industrial equipment, agriculture, mining, construction, metalworking, petroleum chemicals	108,000	Increasing	Expansion may be somewhat slowed in next few years, but increases are still projected
Construction and contractor's products	79,000	Increasing	Market should increase at an above-average rate in next few years
All others, including ordnance, export, and miscellaneous chemicals	51,000	Little change	Some items will increase, others decrease

[a]Derived from National Research Council.[229]

4

Biologic Interactions of Chromium

Chromium is an essential trace element, active in very small concentrations. The toxic effects of excessive concentrations of chromium, as of other micronutrients, have to be considered separately from the effects of biologic doses in which it exerts its essential role.

GENERAL TISSUE INTERACTIONS

In its hexavalent state (as chromic oxide, chromate, or dichromate), chromium is a strong oxidizing agent and readily reacts with organic matter in acidic solution, leading to reduction to the trivalent form.[68,274] The gradual color change of the conventional dichromate cleaning solution from orange to green after exposure to traces of organic material, characteristic of the trivalent state, is well-known. Complexes in which hexavalent chromium would be stabilized against reduction by organic matter are not known. The known toxic action of chromates and of polyoxyacids—e.g., permanganate—of other transition elements is to a large degree due to this oxidizing action. Another property of chromates is the ease with which they penetrate biologic membranes. These properties clearly distinguish hexavalent chromium from the much less toxic trivalent form. The question of the valence state of chromium found in biologic material has not been unequivocally answered.

All the biologic interactions of chromate should result in reduction to the trivalent form and later coordination to organic molecules. This has been demonstrated in the effect of chromate on skin, the interaction of chromates or dichromates with nucleic acids and with wine, and the fate of chromate injected into experimental animals. Earlier studies led to the postulate that the hexavalent state exists in biologic material.[297] However, Feldman found that all the hexavalent chromium injected intravenously into rats was rapidly reduced, as it was if added *in vitro* to various rat tissues.[98]

Divalent chromium probably does not exist in biologic material, inasmuch as it is easily oxidized even by air, but the possibility of its occurrence has been mentioned.[319]

There is some evidence that chromium can exist in other valence states. Special complexes have been prepared in which chromium appears to have ionic valences of 0, 4, or 5 (of which the last two are extremely rare). Examples of the first type are the "sandwich complexes." They are extremely stable and have been mentioned as possibly being involved in the binding of the element to ribonucleic acids.[358] The possibility that these exotic valence states play a role in biologic systems deserves careful examination. At present, there is no evidence of such a role, and it must be concluded on the basis of available data that chromium in biologic matter rapidly attains the trivalent form.[210]

LABELING OF RED BLOOD CELLS

Of particular interest and clinical application is the use of chromate for the labeling of red blood cells. When chromate is added to blood, either *in vitro* or *in vivo,* it penetrates the red-cell membranes without appreciably reacting with components of the plasma.[120] This penetration is rapid: The cells reach 50% chromate saturation *in vitro* within approximately 15 min. Once inside the cells, the chromate is reduced to the trivalent form and bound to hemoglobin, resulting in a stable tagging of the red cells. That the intracellular chromium is trivalent is indicated by its inability to penetrate the red-cell membrane a second time and also by the much greater affinity of trivalent than of hexavalent chromium for a direct *in vitro* reaction with hemoglobin.[54,157] The use of hexavalent chromium makes it possible to reach intracellular concentrations that are toxic to the red cell, probably because of their strong oxidizing action. Glutathione reductase is inhibited by concentrations of

5–25 μg/ml in the cell.[173] The trivalent form, however, binds to the serum proteins and does not penetrate the blood-cell membranes at an appreciable rate.

PROTEIN INTERACTIONS

The best known interaction of chromium with proteins occurs in the tanning process, which conventionally starts with a solution of hexavalent chromium. Tanning is initiated by adding a reducing agent to the chromate solution, which results in the production of trivalent chromium species in a polynuclear complex form. The exact chemistry of the tanning reaction is poorly understood, but it probably involves the coordination of chromium to the carboxyl groups of collagen strands. The result is an extensive cross-linking, via chromium, of the fibrils of the collagen. Chrome tanning results in nearly total saturation of protein with the metal, and various leathers have chromium concentrations of 4–6%.[176,273,323]

At more nearly physiologic concentrations, chromium is known to cross-link protein, without resulting in tanning. This has been studied with conarachin (a protein isolated from groundnuts), in which chromium, but few of the other transition elements, produced a strong, stable cross-linking.[227] Chromium has also been shown to stabilize polypeptide chains in either random-coil or spiral conformation and to prevent thermal transition from one form to the other.[243] Much higher amounts of chromium precipitate protein out of aqueous solution. The reaction is reversible, and no denaturation occurs.[220] Egg protein and human plasma proteins bind trivalent chromium strongly. The hexavalent form reacts with protein only at a low pH and probably through weak bonds that become weaker as physiologic pH is approached.[122] In the intact organism, trace amounts of injected or ingested chromium are bound and carried by the iron-carrying protein, transferrin.[147] This reaction occurs only with the trivalent form; hexavalent chromium would have to be reduced before coordination could occur.

The reaction of hexavalent chromium with skin involves reduction to the trivalent form. Methionine, cystine, lactic acid, hemoglobin, and globulins have been implicated as reductants and potential ligands.[200,281,282] (See Chapter 7 for further discussion of effects of chromium on skin.)

EFFECTS ON ENZYME REACTIONS

Like most other transition elements, chromium can inhibit enzyme reactions when excessive concentrations are added to *in vitro* preparations. Because chromium has been shown to initiate cross-linking of protein at molar chromium : protein ratios of 10 : 1 or 20 : 1,[227] the range of chromium concentrations at which an activating effect of chromium can be expected is narrow. Chromium has been reported to stimulate oxygen consumption in a succinic cytochrome dehydrogenase system,[149] to stimulate phosphoglucomutase,[324] and to stimulate the conversion of acetate to carbon dioxide, cholesterol, and fatty acids in rat liver *in vitro.*[72] These effects are not absolutely specific for chromium and can be regarded as instances of metal activation. However, the digestive enzyme, trypsin, appears to contain chromium as an integral part. The removal of chromium by dialysis is followed by loss of enzyme activity; restoring chromium to the enzyme results in restoration of initial activity. The metal : protein ratio appears to be close to 1 : 1.[180] Chromium was also reported to stimulate the activity of another protein-splitting enzyme, renin.[206] As could be predicted, excessive chromium in either the trivalent or the hexavalent state has been found to inhibit bacterial urease activity[138] (at 1–10 μg of chromate per milliliter), thromboplastic activity[57] (at 1.25 mg/200 μg of protein), and beta-glucuronidase activity[101] (no exact data are available).

INTERACTIONS WITH NUCLEIC ACIDS

Hexavalent chromium reacts with nucleic acids even in tissue slices; it greatly reduces the amount of nucleic acids extractable with trichloroacetic acid.[142] Because the effect is specific for chromates and is not observed with other acids and because the treatment imparts a greenish color to the tissue, it is likely that the reaction includes a reduction to trivalent chromium and later complex formation with the nucleic acid.

Nucleic acids contain high concentrations of chromium, as well as other metals. Chromium concentrations ranging from 260 to more than 1,000 μg/g have been detected in a beef liver fraction consisting of 70% RNA and 30% protein, and even highly purified RNA fractions contain 50–140 μg/g.[357,358] The exact function of chromium and of the other trace elements in RNA is unknown. But it has been shown that, *in vitro,* transition elements (including chromium) stabilize the ordered structure of RNA.[110] Whether this is also true *in vivo* is still a matter of conjecture. It is safe to assume that chromium is linked to the components of nucleic acid by coordination; however, it is not known whether trivalent chromium is involved or whether the bond involves chromium of zero valence, owing to formation of sandwich complexes.[358]

EFFECTS ON BACTERIA

The effects of chromium on bacteria depend on its concentration, but also on a variety of other factors, such as the bacteria themselves and the composition of the medium. In general, concentrations approaching 1 μg/ml of medium are toxic, particularly in the case of hexavalent chromium. This has been shown with *Corynebacterium diphtheriae,*[61] *Staphylococcus aureus,*[138] phage T, and *Escherichia coli.*[260] Mutation can increase sensitivity to chromium, as shown in *Salmonella typhimurium.*[67] However, some organisms reportedly respond to chromium with an increased function—for example, *Aerobacter aerogenes,*[245] Bacillus,[361] *Propionibacterium shermanni,*[254] some chlorophyceae,[143] and diatoms.[143] Most of the effects observed are not specific for chromium and can be seen with other elements. It appears that an absolute requirement for chromium has not yet been defined for bacteria.

YEAST INTERACTIONS

The stimulation of fermentation observed in yeast when chromate at 125–200 μg/ml is added is probably a toxic effect and not specific for chromium, inasmuch as it can be simulated by a number of other nonspecific stimuli.[301] Even when supplied in the trivalent form at 200 μg/ml, chromium is toxic to brewer's yeast grown in a synthetic medium.[136] However, addition of trivalent chromium at 0.1 μg/ml to a culture of *Saccharomyces carlsbergensis* grown in Sabauraud's medium resulted in a significant stimulation of carbon dioxide production, after a lag phase of 3 hr.[47]

Brewer's yeast, but not torula (*Torulopsis utilis*) yeast, is an excellent source of available chromium.[210] Cellular uptake of radioactive chromium salts from the medium is greatly enhanced by the addition of glucose to the medium.[47]

The form in which chromium occurs in yeast has not been identified; it is probably a low-molecular-weight organic complex, water-soluble and heat-stable. The complex is named "glucose tolerance factor"; it can be extracted from yeast, and its biologic effects can be measured in various systems. Addition of this factor to chromium-deficient yeast causes an immediate increase in fermentation, without the lag phase seen when inorganic salts are added.[47] The intestinal absorption of this factor in rats is much greater than that of simple chromium salts, and it is transported from the maternal organism into the fetus against a gradient, whereas simple chromium salts do not cross the placenta at all. Its effects on glucose metabolism are much stronger than, but qualitatively similar to, those of simple salts.[210]

SUMMARY

Chromium occurs in most biologic material in the trivalent form, in which it is strongly associated with proteins, nucleic acids, and a variety of low-molecular-weight ligands. Its concentration ranges from a low of a few nanograms per gram in blood plasma to over 1 mg/g in some liver fractions. Nondietary exposures from the environment via air and water furnish a significant proportion of the hexavalent form. The latter, if soluble, will react with components of living tissue and be reduced to the trivalent form. The hexavalent form is more toxic than the trivalent because of its oxidizing potential and its easy permeation of biologic membranes. Excessive concentrations of chromium (micrograms per gram) are toxic in enzyme and bacterial systems. Much lower concentrations have been shown to have stimulatory effects, but many of these were not specific for chromium.

5

Chromium in Nutrition

NORMAL DIETARY INTAKE

Daily chromium intake in the United States varies greatly, depending on dietary preference (see Tables 2-2 through 2-5). The average intake of subjects eating one institutional diet was 52 μg.[189] Another institutional diet furnished 78 μg.[297] The average *ad libitum* intake of students was 65 μg/day, with a range of 5–115 μg/day.[189] The chromium intake in other countries may be considerably higher.[210] Most meats are good sources of chromium; fish, many vegetables, and refined sugar are poor sources.[297,335] The urinary excretion of chromium, the main route of excretion of absorbed chromium, is estimated to be between 7 and 10 μg/day.[156,249,356] This amount is easily covered by most dietary intakes in the United States; but this does not necessarily mean that the dietary intake is sufficient. Chromium is generally poorly absorbed, and its availability varies greatly with the dietary source–ranging from less than 1% to approximately 25% of an oral dose.[211]

EXCESSIVE INTAKE

It appears that the range of toxic concentrations of hexavalent chromium in marine animals begins at 1 mg/liter of water, but the actual toxic concentration depends on the sensitivity of the methods used and on the species. For example, nereid marine worms, rainbow trout, and carcinides (swimming crabs) showed toxic symptoms at water concentrations of chromium of 1, 2.5, and 50 mg/liter, respectively.[262]

Chronic toxicity can be observed in several mammalian species with hexavalent chromium in the drinking water in concentrations of more than 5 mg/liter. At this concentration, the element was found to accumulate in rats, but it caused no changes in growth rate, food intake, or results of blood analysis.[49,198] Even 25 mg/liter in the drinking water failed to produce changes in these characteristics or in the histologic appearance of the tissues after 6 months. Dogs tolerated hexavalent chromium in the water at up to 11.2 mg/liter for 4 years without ill effects.[11] The minimal lethal dose in dogs is approximately 75 mg of chromium as sodium chromate, when injected intravenously.[331] The salt causes acute hypertension, hypocholesterolemia, and hypoglycemia. Growing chickens showed no detrimental symptoms when they were fed 100 μg/g in the diet.[275]

Symptoms of excessive dietary intake of chromium in man are unknown; the known symptomatology is restricted to exposure to airborne chromium or to chromium applied directly to the skin. The United States drinking-water standard rejects water containing more than 0.05 mg of chromium (as hexavalent chromium) per liter.[259] This concentration is almost never reached in domestic water supplies in the United States. However, one family reportedly used drinking water containing 1–25 mg/liter for several years without symptoms of toxicity.[77]

Trivalent chromium is considerably less toxic than hexavalent; its toxicity appears to be restricted to parenteral administration. In mice, the minimal lethal dosage on intravenous injection of trivalent chromium was reported as 2.3 and 0.8 g/kg of body weight for the acetate and chloride, respectively.[331] In rats, the lethal dosage on intravenous injection for 50% of the animals was 10–18 mg/kg for chrome alum and chromium hexaurea chloride, respectively.[215]

Feeding 50–1,000 mg of trivalent chromium complexes per day to cats for 1–3 months had no bad effects.[3] Rats tolerated chromic lactate in the diet at 1–100 mg/kg.[66] The therapeutic: toxic dose ratio for intravenously injected trivalent chromium has been estimated to be approximately 1 : 10,000.[210]

DEFICIENCY STATES

Animals

Mild chromium deficiency has been produced experimentally in rats,[303] mice,[299] and squirrel monkeys.[78] Rats raised on a torula yeast ration low in available chromium developed, as the first symptom, an impairment of tolerance to intravenous glucose. Glucose removal rates (the rate of disappearance of excess glucose from the blood in percent per minute) declined within a few weeks to approximately half their original values.[218] This defect could be prevented by feeding chromium compounds in the diet; it could also be cured overnight by one dose (by stomach tube) of 20–50 μg of chromium in the form of suitable compounds. The poor absorption of these compounds is demonstrated by the finding that less than 1 μg injected intravenously produced the same effect.[215] Fasting hyperglycemia has not been observed in mild chromium deficiency.

The defect in glucose tolerance is caused by a diminished response of the peripheral tissue to the animal's own insulin. The glucose uptake from the medium by epididymal fat tissue *in vitro* is identical in tissues from chromium-deficient and chromium-supplemented animals, if no insulin is added to the incubation medium. However, the addition of insulin results in a much smaller increment of glucose uptake by chromium-deficient tissue than by chromium-supplemented tissue.[216] The findings were identical when glucose utilization for oxidation, for lipogenesis, or for glycogen formation was measured.[212,215] These findings suggested a site of chromium action very close to the first steps of glucose metabolism. Because it could be shown that the cell transport of a nonutilizable sugar, D-galactose, was influenced by chromium deficiency and resupplementation in the same way as in the other systems, it was postulated that chromium acts as a cofactor for the reaction of insulin with the cell membrane.[213] Later studies revealed a universal relation between chromium and insulin: The effect of the hormone by itself in all insulin-responsive systems is less when chromium-deficient donor animals are used, and this effect can be significantly increased by either *in vivo* administration of chromium compounds to the donor animal or *in vitro* addition of chromium to the system. This has been shown for the action of insulin on glucose utilization by the isolated rat lens,[96] for the effect of insulin on water uptake by isolated mitochondria,[53] for the effect of insulin on cell transport of nonutilizable amino acids,[272] for the effect of insulin on the utilization of intracellular amino acids for protein synthesis,[272] and for the disulfide interchange between insulin and membrane sulfhydryl receptors.[59]

Squirrel monkeys fed a commercial laboratory chow low in available chromium also developed impaired glucose and tolbutamide tolerance. The impairment can be prevented by adding chromium to the drinking water at 10 μg/liter. The normalization of impaired glucose tolerance with chromium supplements was slow; it took approximately 22 weeks.[78]

As stated earlier, impairment of glucose tolerance is the first response of animals to a mild chromium deficiency. A more severe deficiency can be produced by raising animals in an environment that allows strict control of airborne contamination or by subjecting them to additional stress, such as low-protein diets, hemorrhage, or strenuous exercise. The mortality of male mice raised in an environment allowing maximal exclusion of trace contamination was decreased if chromium was added to their drinking water at 5 mg/liter. In the case of rats, early mortality was not significantly affected, but chromium supplementation increased the lifespan of survivors: The mean age of the last surviving 10% was increased from 1,141 to 1,249 days. Most mice and rats grew significantly better with chromium supplementation than their deficient controls.[298,299] Chromium deficiency in these rats also resulted in significantly increased concentrations of circulating cholesterol; this increase was prevented by dietary chromium. Chromium-supplemented rats had a significantly lower incidence of aortic plaques than the chromium-deficient controls.[296]

Fasting hyperglycemia and glycosuria were observed in rats raised under maximal exclusion of chromium contamination. More than half the 185 chromium-deficient rats showed a positive test for urine sugar, compared with only nine of 87 chromium-supplemented controls.[293]

The combination of chromium deficiency and protein deficiency, produced by feeding rats a diet containing only 10% of isolated soy protein, results in degeneration of the cornea, manifested by vascular infiltration and opacities.[271] These are completely prevented, but not reversed, by dietary chromium supplementation. Further increasing stress to the animals by the controlled withdrawal of a measured volume of blood resulted in 67% survival of the deficient rats versus 92% of the controls in one experiment and 27% versus 60%, respectively, in another.[212] The sequence of events as the severity of chromium deficiency increases is (1) mild impairment of glucose tolerance, (2) mild impairment of growth and longevity, (3) fasting hyperglycemia and glycosuria, and (4) diminished resistance against stress. Impaired glucose tolerance is relatively easy to observe in experimental animals; the latter three conditions require careful control of environment and diet.

Man

Chromium deficiency in man can be diagnosed only by a combination of criteria:

1. decline of chromium concentration in an indicator tissue from a

pre-established normal value, which is reversible by supplementation with chromium at a nutritionally reasonable rate;

2. impairment of a physiologic response known to occur in normal persons and reconstitution to normal after chromium supplementation; and

3. impairment of a physiologic function known to depend on chromium and reconstitution to normal after chromium supplementation.

Ideally, these three criteria should be combined for the diagnosis of chromium deficiency in a given person. However, that has not yet been possible, and the preliminary diagnosis of chromium deficiency rests mainly on the demonstration of a chromium-responsive impairment of physiologic function.

ANALYTIC FINDINGS

The chromium concentration in most tissues sampled at autopsy appears to decline with increasing age in the population of the United States.[297,332] Chromium is the only element for which this behavior has been observed.[332] Repeated pregnancies result in a significant decrease in hair chromium in the mother, which suggests that gestation imposes a chromium demand that is not easily covered by dietary intake.[133] Hair chromium content increases in the fetus with gestational age; premature infants are born with a low hair chromium content.[130]

The concentration of chromium in blood, serum, or plasma in the fasting state is not correlated with the nutritional chromium status.[130,210] However, it has been found that the acute increase in plasma chromium concentration after a challenge with glucose or insulin can be used as a criterion of chromium nutrition, in spite of considerable analytic difficulties.[112] Young, healthy subjects respond to a glucose or insulin load with an acute rise in plasma chromium concentration. Elderly subjects and those with impaired glucose tolerance generally do not experience this increase. It has been shown that increasing the dietary chromium intake over a period of several weeks can restore the plasma chromium response.[112] Similar findings have been obtained in chromium-deficient rats.[211]

Recent studies suggest that the decreased chemotactic index of polymorphonuclear leukocytes in some insulin-dependent diabetics and in gestational diabetics is another symptom of chromium deficiency. The chemotaxis can be increased to normal by incubating the leukocytes in a 25-ng/ml solution of trivalent chromium (as chromium chloride). In the absence of chromium, high concentrations (100 μU/ml) of exogenous insulin are required to obtain a similar effect.[132] The combination of low hair chromium, low urinary chromium, and the absence of a plasma chromium response to glucose is a strong indication of a suboptimal chromium nutritional state.[130] The final proof of the existence of chromium deficiency, however, can come only from therapeutic trials.

CHROMIUM SUPPLEMENTS

Impairment of glucose tolerance is the first symptom of chromium deficiency in experimental animals. Glucose intolerance is very common in man.[326] Its possible causes are many, and it is possible that chromium deficiency is one of them.

A study performed in a metabolic ward with four diabetic subjects resulted in significant improvement of oral glucose tolerance in three after several weeks of supplementation with daily doses of 180–1,000 μg of chromium. Of two diabetics observed as outpatients for an extended period, one was improved and the other was unaffected by chromium. Chromium did not improve glucose metabolism in any of those diabetics to such a degree that their hypoglycemic medication could be reduced; therefore, chromium supplementation was not considered to be a therapeutic agent in diabetes.[113]

Supplementation with 150 μg of chromium per day was ineffective in two double-blind studies. One was conducted for 6 weeks with 16 diabetic outpatients,[295] the other with prison inmates.[307]

In another study investigating the effect of chromium on the impaired glucose tolerance of nondiabetic elderly people, four of 10 responded to prolonged chromium supplementation (150 μg/day) with significant improvement or normalization of glucose tolerance.[189]

A more severe and responsive chromium deficiency appears to exist in some areas of the world in connection with protein–calorie malnutrition. The syndrome of kwashiorkor and marasmus often includes marked impairment of glucose tolerance, which slowly improves when the children are fed a therapeutic high-protein diet. In a study performed in Jordan, the administration of one dose of 250 μg of chromium to each of six children with impaired glucose tolerance resulted in an immediate and highly significant improvement of the test in every child.[146] A similar study was performed on six malnourished children in Nigeria; again, chromium supplementation improved glucose tolerance significantly.[146] Repeated glucose tolerance tests in five children, 3 days after the initial test but without chromium supplement, did not reveal any significant changes. A third study, performed in Turkey, had essentially the same results.[126] However, malnourished children with impaired glucose tolerance in Egypt failed to respond to chromium supplementation. These children had a high previous chromium intake and blood chromium content, which suggested that their chromium status was normal and their impaired glucose tolerance was not due to chromium deficiency.[55]

These studies show clearly that not all forms of impaired glucose tolerance are due to chromium deficiency, but they suggest that, in some areas and in some age groups, chromium nutrition is suboptimal.

IS CHROMIUM AN ESSENTIAL ELEMENT?

Chromium deficiency can be produced in experimental animals; it can be prevented and cured by appropriate chromium supplementation. Its symptoms are reproducible and consist of a general decrease in the tissue response to insulin. On this basis, chromium must be considered an essential element. It is likely that suboptimal nutritional states do exist with regard to chromium in several populations, but the extent of chromium deficiency is unknown at present.

The widely varying availability for absorption of different chromium compounds makes it very difficult to assess the nutritional situation within the United States. To compensate for the average daily loss of chromium in the urine, amounting to approximately 10 μg, a daily intake of 40 μg would be necessary if it were furnished in the best available form—for example, as glucose tolerance factor from yeast (see Chapter 4). If supplied in the form of inorganic salts, which are poorly available, 2,000 μg would be necessary. It has been shown that the availability of chromium in different foods ranges between these extremes. The estimated average intake of approximately 60 μg would be sufficient if the availability were high; this is clearly not the case. Therefore, marginal deficiency of chromium is of greater concern to nutritionists than overexposure.[129]

SUMMARY

Chromium must be considered as an essential element. Its deficiency results in impaired glucose metabolism due to poor effectiveness of insulin. Chromium-responsive forms of impaired glucose tolerance are known to exist in the United States, particularly in elderly people. The intake of chromium in the United States is considered marginal.

6

Absorption, Metabolism, and Excretion of Chromium

Before the quantitative aspects of chromium metabolism are discussed, it must be emphasized that the analytic chemistry related to ultratrace amounts of chromium in biologic material is still in a state of development and that the reported concentrations of chromium in such material are still being revised downward.[130,244] Thus, although the existing analytic data allow a qualitative description of chromium metabolism, quantitative statements must be interpreted with caution.

ABSORPTION

The mechanism by which chromium is absorbed from the gastrointestinal tract is poorly understood. The degree of absorption depends strongly on the chemical form in which the chromium is bound and ranges from a reported 0.1–1.2% of trivalent chromium salts to approximately 25% of glucose tolerance factor (see Chapter 4). The presence of some ligands, such as amino acids, and some chelating agents significantly increases chromium transport *in vitro*.[211] Tipton *et al.* have reported positive chromium balance in four subjects who consumed 200–400 μg of chromium, of which up to 50% appeared in the urine, thus suggesting 50% absorption.[333,334] These data contrast with the much lower dietary and urinary concentrations detected by other investigators.[130,156,189,356] An exact value for intestinal absorption can therefore not be given.

The problem of chromium absorption from the air is at least as complicated. Chromium concentrations in the lungs are consistently among the highest in the body. There are marked geographic variations—for example, a range of 90–720 μg/g of weight for the San Francisco, New York, and Chicago areas.[297] These variations appear to be loosely correlated with the chromium content of the air. It is unlikely that the intake from air under ordinary conditions contributes significantly to the total intake of available chromium; the intake from air is calculated to be less than 1 μg/day. But excessive exposure to airborne chromium does result in increased intake, as evidenced by increased chromium content of blood and urine.[19,356]

The mechanism of absorption of airborne chromium is poorly understood, but there appear to be three pathways:

1. Airborne chromium with a particle size greater than 1 μm is believed not to reach the alveoli at all, but to be trapped in the bronchi.[292] These particles are later moved to the pharynx by ciliary action and swallowed. This chromium could be counted as part of the dietary intake.
2. Particles small enough to penetrate into the alveoli may be trapped in the tissue if insoluble.
3. If they can be dissolved, particles may be able to penetrate into the blood and distribute themselves throughout the body.

Chromium concentrations in the lung do not decline with age, as they do in most other organs, but tend to increase; that suggests strongly that most of the chromium in the lungs is not in equilibrium with that in the rest of the body. It is likely, but not proved, that relatively inert forms of chromium, such as oxides and hydroxides of trivalent chromium, would stay in the lung tissue, whereas the more soluble forms, such as chromates, would penetrate into the bloodstream. It is difficult to assess this situation, because most air chromium measurements do not distinguish between valence states and only a few are concerned with particle size.

METABOLISM

Transport

In the rat, chromium absorbed by the intestines is almost entirely bound to transferrin, the iron-carrying protein;[147] chromium in excessive doses is also bound to other protein fractions. However, in man, the bulk of an administered dose of radioactive chromium chloride is carried in the albumin fraction, and only 30–40% in globulins, of which transferrin is a part.[83] Chromium disappears rapidly from the blood and is taken up by other tissues, where it is concentrated much more heavily—by a factor of 10–100—than in the blood. This indicates that the blood ordinarily is not a usable indicator of chromium nutritional status.[210,211]

However, a very useful measure of chromium can be obtained from the blood plasma. After a challenge with a glucose load or insulin, the plasma chromium concentration has been found to rise rapidly to approximately twice the concentration in the fasting state. This phenomenon occurs only in healthy young people, and the increase becomes progressively less with increasing age or with disturbances of glucose tolerance. Chromium supplementation prolonged over several weeks has been shown to restore the acute plasma response to normal, which suggests that an existing deficiency of biologically active, available chromium had been corrected.[112] The site of the specific pool from which this chromium is derived has not yet been identified, but it is clear that it is not accessible to simple chromium salts in short-term experiments. It can be labeled in rats, however, by the administration of the organic chromium complex, glucose tolerance factor.[211] Plasma chromium concentrations usually return to their starting value within 2–3 hr of the glucose or insulin challenge; much of the increased chromium is excreted in the urine.[295]

Distribution

The distribution of a test dose of chromium depends on its chemical state and the amount administered. Many studies have used amounts far beyond the physiologic range. It is therefore not completely clear whether the great affinity of the reticuloendothelial system, liver, spleen, and bone marrow for chromium is a physiologic phenomenon or an expression of an overload.

With low tracer doses of chromium chloride, uptake in rats was highest in ovaries and spleen, and next highest in kidney and liver; lower concentrations were found in lung, heart, pancreas, and brain.[145] The distribution of chromium administered to rats in the form of glucose tolerance factor is entirely different, with liver accumulating the highest concentration, followed by uterus, kidney, and bone; lung, heart, intestines, spleen, pancreas, and brain accumulated about 50% of the concentration found in liver; muscle, ovaries, and aorta had a considerably lower affinity.[211]

Those findings in experimental animals contrast clearly with results of human autopsies. In man, the highest concentration of chromium is in the lung, which suggests that the human lung obtains most of its chromium from exposure to air, and not from ingested chromium. It is

TABLE 6-1 Distribution of Chromium in Human Tissues

Sample[a]	Reference	Chromium Concentration, ng/g or ng/ml
Unexposed Subjects		
Dried blood	367	5,000
Serum	297	520
Serum	297	170
Plasma, serum	99,141,189	20–30
Serum	130	8
Serum	83	2
Urine, adults, 24-hr excretion	130	8.4 μg (total)
Urine, children, 24-hr excretion	130	5.5
	249	4
	156	5
	297	860
Hair		
newborn	130	974
maternal	130	382
Lung		
New York–Chicago	297	720
Denver	297	140
Liver		
New York–Chicago	297	270
Denver	297	30
Kidney		
New York–Chicago	297	90
Denver	297	40
Heart		
New York–Chicago	297	130
Denver	297	30
Exposed Subjects		
Blood cells, chromate workers	19	30, 54, 140
Plasma, chromate workers	19	0, 20, 17
Blood, chromate workers	97	40–60
Urine, chromate workers	19	52, 0, 3.6, 0
Urine, chromate workers	97	43–71
Urine, chromate workers	356	1,600
Urine, chromate workers	202	34–121
Lung, chromate workers (acid-soluble Cr)	19	2,800–161,000[b]

[a]Only a few examples from the literature are given to show the wide range of reported concentrations. For blood, urine, and hair, the lowest values reported probably are the most reliable ones.
[b]Dry-tissue basis.

of particular interest that almost half the total cellular chromium in liver was found to be concentrated in the nuclei. This distinguishes chromium from iron, zinc, copper, molybdenum, and manganese[90] and may be related to the fact that the testis accumulates chromium very rapidly after injection of a tracer dose and that a similar increase occurs in epididymis when the testicular concentration begins to decline, which suggests incorporation of chromium into sperm.[145]

The values reported in Table 6-1 must be interpreted with caution. Chromium analysis, particularly for very low concentrations, such as those in blood and urine, has not yet been well standardized. And chemical analysis gives no information on the biologically significant fraction of the total chromium content.

EXCRETION

Chromium is excreted in urine and feces. The urinary pathway is the major one, accounting for at least 80% of injected chromium. Several investigators have detected a fecal excretion of intravenously injected chromium ranging from 0.5 to 20% of a given dose.[83] Because, in using experimental animals, it is difficult to obtain feces uncontaminated by urine, the higher figures must be interpreted with caution. Nearly all chromium in urine is present in the form of low-molecular-weight complexes; protein-bound chromium is excreted to only a very small degree. Chromium that is present in the blood coordinated to small molecular ligands is filtered at the glomerulus, and up to 63% is reabsorbed from the filtrate in the tubules.[65]

There is little agreement on the amount of daily urinary chromium excretion, and concentrations from nondetectable[130] to 860 μg/liter[297] have been described. In two carefully controlled studies, averages of 5 and 4 μg/liter were found.[156,249] Mean 24-hr urinary chromium excretion of 20 young adults has been given as 8.4 μg, with a range of 1.6–21 μg.[130] These values can probably be accepted with confidence. Very high chromium excretion appears to occur in some insulin-dependent diabetics.[130]

The excretion of injected trivalent chromium, as measured by whole-body counting, has an exponential character. There are three main components, with half-lives of 0.5, 5.9, and 83.4 days. In experimental rats, this behavior of chromium chloride is not influenced by dose or nutritional state; hence, it might not be representative of the biologically important fraction of chromium, which may be a small part of the total.[215] Exact data on the biologic half-life of chromium in man are limited. It appears from the results of therapeutic trials that chromium turnover is very slow. On termination of chromium supplementation, the glucose tolerance test usually remains normal for approximately 30 days and then slowly returns to its impaired value. If the three-component excretion kinetics found in experimental animals is valid in man, one can assume that the first, rapid component (half-life, 0.5 days) is an effective means of eliminating excessive chromium rapidly and efficiently.

PLACENTAL TRANSPORT

In spite of considerable efforts by several investigators, no transfer of chromium from mother to fetus, regardless of valence state or chemical form, has been demonstrated.[355] Yet chromium can consistently be found in the newborn and in the fetus, if sensitive methods are used.[221] The chromium concentration in the newborn rat cannot be increased by feeding the mother chromium chloride, even at very high concentrations, in the drinking water. It is, however, influenced by natural sources of chromium in the diet. Feeding mothers a natural diet rich in chromium resulted in more than doubling the chromium concentration in the young, compared with the offspring of mothers fed a chromium-deficient torula yeast diet.[214] It was later demonstrated that chromium is transported into the fetus only if it is present in a special organically bound form, glucose tolerance factor. Administration to pregnant rats of extracts obtained from brewer's yeast that was grown in a medium containing chromium-51 resulted in high fetal chromium concentrations.[214] Comparing the chromium in fetal organs with that in maternal organs showed that glucose tolerance factor chromium was concentrated in the fetus against a gradient: Fetal concentrations exceeded maternal.[211] On the basis of these findings, it appears that chromium must be present in a special organic complex if it is to be available to the fetus. Inasmuch as the mother does not appreciably synthesize this complex from inorganic chromium, the glucose tolerance factor assumes the character almost of a vitamin.

SUMMARY

The metabolism of chromium depends heavily on its chemical form. Animal experiments suggest that one or more specific organic chromium complexes, designated "glucose tolerance factor," are handled by an organism in such a way as to meet the criteria for an essential element. The simple chromium salts are poorly absorbed and not transported across the placenta, and homeostatic mechanisms are difficult to detect. Data on tissue distribution of chromium are difficult to interpret, because present methods do not distinguish between contaminating chromium from the environment and glucose tolerance factor, the specific chromium complex that seems to be physiologically active.

7

Effects of Chromium Compounds on Human Health

Chromium as a metal is biologically inert and does not produce toxic or other harmful effects in man or laboratory animals. Compounds of chromium in the trivalent state have no established toxicity. When taken by mouth, they do not give rise to local or systemic effects and are poorly absorbed. No specific effects are known to result from inhalation. In contact with the skin, they combine with proteins in the superficial layers, but do not cause ulceration. Their possible role in dermatitis is discussed in the latter part of this chapter.

The chief health problems associated with chromium are related to hexavalent chromium compounds, which are irritant and corrosive and may be absorbed by ingestion, through the skin, and by inhalation. Acute systemic poisoning is rare and usually accidental. It may follow accidental or deliberate ingestion or result from absorption through the skin. Brieger[44] reported the death of 12 persons after application of an antiscabietic ointment in which sulfur had been replaced with hexavalent chromium. Necrosis of the skin occurred at sites of application and was followed or accompanied by nausea, vomiting, shock, and coma. The patient's urine contained albumin and blood. The main postmortem findings were tubular necrosis and hyperemia of the kidneys. A number of other cases of kidney damage have been reported in connection with attempted suicide or therapeutic application.[44,116,199,240,261]

Knowledge of the harmful effects of hexavalent chromium is derived almost entirely from occupational exposures, where the effects are on the respiratory system and skin. These effects are described in detail in this chapter.

RESPIRATORY EFFECTS

Ulceration and Perforation of the Nasal Septum

Chromium in hexavalent form is clearly a cause of ulceration of the skin and of ulceration and perforation of the nasal septum. The septum is particularly susceptible to the action of chromium, not only because of the immediate contact of inhaled particles with the septum, but also because of its structure. The cartilaginous framework of the nose consists of five pieces—the two upper and the two lower lateral cartilages and the cartilage of the septum. The perforation is limited to the cartilage of the septum. This is because the mucous membrane covering this area is far less vascular than the mucous membrane lining the rest of the nasal fossae and thus is easily destroyed. This destruction cuts off the blood supply to the cartilage, and necrosis results. The ulcerative process, after having progressed upward as far as the junction of the septum with the ethmoid and backward to the vomer, is then arrested—the bone is not attacked. Healing then takes place, and the cicatrix usually becomes covered with an ecthymatous crust of mucus. In no instance is the anterior of the lower border of the septum destroyed; consequently, the rigidity of the parts is maintained, and deformity is absent. The beginning of the process is characterized by sneezing and the ordinary symptoms of nasal catarrh. The pain accompanying the ulceration appears to be insignificant—at any rate, it is usually not severe enough to necessitate absence from work or to warrant medical treatment. The only apparent inconvenience is related to the formation of mucous plugs in the nasal passages.

Other sites where ulcers may appear as a result of contact with chromium are the skin, the roots of the fingernails, the knuckles, the eyelids, the edges of the nostrils, the toes (if a person's shoes become soaked with chromic acid or chromate), and, rarely, the throat. The ulceration is generally slow and not very painful.

The ulcerative property of hexavalent chromium was first described in 1827 (with regard to cutaneous lesions) by William Cumin,[70] a physician in Scotland. In France, Bécourt and Chevallier[24] in 1863 and Delpech and Hillairet[79] in 1869 reported both cutaneous and mucous-mem-

TABLE 7-1 Clinical Findings in Workers Employed in Chromium-Plating Plants[a]

Case	Occupation	Time Employed in Chromium-Plating Room, mo	Time Over Tank, hr/day	Approximate CrO_3 Exposure, mg/m^3	Perforated Septum[b]	Ulcerated Septum[b]	Inflamed Mucosa[b]	Nosebleed	Chrome Holes
1	Chromium plater	6	4	1.5	++	–	++	Yes	Yes
2	Chromium plater	20	4	2.8	++	–	+	Yes	Yes
3	Foreman plater	7	2	2.5	–	++	++	Yes	No
4	Foreman plater	8.5	3	2.5	–	++	++	Yes	No
5	Chromium plater	3.5	4	5.6	–	++	++	Yes	Yes
6	Chromium plater	0.75	7	0.12	–	–	++	Yes	Yes
7	Chromium plater	0.25	7	0.12	–	–	++	Yes	No
8	Chromium plater	7	7	0.12	–	–	++	Yes	No
9	Chromium plater	3	7	0.12	–	–	++	No	Yes
10	Chromium plater	36	4	0.2	–	–	++	No	No
11	Chromium plater	5	6	0.12	–	–	+	Yes	Yes
12	Chromium plater	0.75	6	0.12	–	–	+	No	No
13	Chromium plater[c]	12	4	2.8	–	–	–	No	No
14	Chromium plater	0.67	2	2.8	–	–	–	No	No
15	Nickel plater[d]	1.5	0	?	–	+	+	Yes	No
16	Racker	8	0	?	+	–	+	Yes	No
17	Racker	0.75	0	?	–	–	+	No	No
18	Racker	0.75	0	?	–	–	+	No	No
19	Wiper	1.5	0	?	–	–	+	No	No
20	Foreman[e]	0	0	0	–	–	+	No	No
21	Foreman[e]	0	0	0	–	–	+	No	No
22	Clerk[e]	0	0	0	–	–	–	No	No
23	Inspector[e]	0	0	0	–	–	+	No	No

[a]Derived from Bloomfield and Blum.[39]
[b]++, marked; +, slight; –, negative.
[c]Used vaseline in nose.
[d]Cyanide burns.
[e]Worked in other departments of factory.

brane effects observed among bichromate workers. In 1851, apparently without any knowledge of Cumin's paper, Bécourt and Chevallier had noted what was to them a new disease of the skin, occurring in some chromium works near Paris.[24] They had written to physicians in other countries asking whether such things had ever been seen elsewhere and received an answer from Isaac of Baltimore, in 1852, confirming their observations and noting that Maryland chrome workmen protected themselves from nasal damage by tying a wet sponge over nose and mouth.

Wutzdorff says that similar lesions were found in 1889 by German factory inspectors among workers in the newly opened chromate works in Griesheim.[370] Leymann examined 722 workmen who were in contact with chromates and found ulcers and perforation of the septum in 253 (35%).[261]

In 1911, K. B. Lehmann made a thorough study of the use of chromates in German industry.[185] He felt these compounds to be comparatively harmless. The poisoning that did occur was generally local; systemic symptoms were extremely rare. Even the local lesions were neither frequent nor severe.

In 1902, in England, Legge found perforation of the nasal septum in 72% of 176 chrome workmen and ulceration without perforation in 11%.[184] Some workers developed ulcers after only 2 months of employment. The absence of perforation in some workers probably was due to a less intense exposure.

In 1928, Bloomfield and Blum published a report on physical examinations of 23 men employed in six plants engaged in chromium plating in the United States.[39] In these same plants, they collected by an impinger method 39 air samples from the chrome-plating room for chromic acid analysis. Nearly all samples were taken from above the chromium-plating tanks, near the point where the operator stood and at breathing level. Of the 23 men examined, four were never in the plating room and were selected as controls. Five other workers examined were not chrome platers, but worked in the plating room, about 10 ft or more from the tank. The remaining 14 men worked near or over the plating tanks. The 39 air samples were intended not primarily to be related to symptoms in the workers, but apparently to see how varying ventilation over the tanks might affect chromic acid concentrations. The authors did report, however, an estimated chromium trioxide exposure for each of the 23 men examined, as well as their symptoms. The complete results are shown in Table 7-1. Except for technical details regarding air sampling and measurements, the paper contains no relevant data. The authors concluded that "continuous daily exposure to concentrations of chromic acid greater than 1 mg/10 m^3 [0.1 mg/m^3] is likely to cause a definite injury to the nasal tissues of the operators." Inasmuch as no concentrations lower than 0.12 mg/m^3 were observed in the chromium-plating room, injury to nasal tissues caused by lower concentrations could not be ruled out.

In 1965, Kleinfeld and Rosso reported on examinations of nine workers in a chrome-plating plant, with results as shown in Table 7-2.[167] Analysis of an unreported number of air samples showed concentrations of chromium of 0.18–1.4 mg/m^3. Seven of the nine men had some degree of septal ulceration; four of the seven had septal perforation. As in the study by Bloomfield and Blum, concentrations below 0.1 mg/m^3 were not studied, nor were the effects of time at a fixed concentration.

The United States Public Health Service conducted a study on the health of 897 workers in seven chromate-producing plants in the early 1950's. The results are shown in Table 7-3. Nasal perforation was considerably more prevalent in nonwhite than in white workers, and its prevalence increased with time worked in the chromate industry. Baetjer (personal communication) believed that the earlier appearance and higher rate in nonwhite workers was due to their employment in areas of greater exposure.

In the Public Health Service study, a total of about 1,800 samples were taken to measure the atmospheric environment, but the results of physical examination were not related to chromium exposures, so data similar to those published by Bloomfield and Blum are not available. Thus, although the Public Health Service study is probably the most extensive ever undertaken in the chromate industry, it is of only limited use in estimating safe limits of exposure.

TABLE 7-2 Nasal Medical Findings in a Chromium-Plating Plant[a]

Case	Age, yr	Duration of Exposure, mo	Findings
1	30	6	Perforated septum
2	19	2	Perforated septum
3	19	12	Perforated septum
4	18	9	Perforated septum
5	47	10	Ulcerated septum
6	45	6	Ulcerated septum
7	23	1	Ulcerated septum
8	20	0.5	Moderate injection of septum and turbinates
9	48	9	Moderate injection of septum

[a]Derived from Kleinfeld and Rosso.[167]

TABLE 7-3 Perforation of Nasal Septum in Chromate Workers[a]

	All Workers			White Workers			Nonwhite Workers		
Time Worked in Chromate Industry	Total No.	Workers with Perforation No.	Workers with Perforation %	Total No.	Workers with Perforation No.	Workers with Perforation %	Total No.	Workers with Perforation No.	Workers with Perforation %
Less than 6 months	41	1	2.4	32	0	0	9	1	11.1
6 months–3 years	117	46	39.3	89	28	31.5	28	18	64.3
3–10 years	370	205	55.4	235	104	44.3	135	101	74.8
10 years or over	369	257	69.6	297	190	64.0	72	67	93.1
TOTAL	897	509	56.7	653	322	49.3	244	187	76.6

[a]Derived from Federal Security Agency.[97]

Mancuso[202] reported on physical examinations of a random sample of 97 workers from a chromate-chemical plant and related the prevalence of nasal septal perforation to exposures to trivalent and hexavalent chromium. The results are shown in Table 11-2 (p. 92). It is not stated whether any of these exposures were below 0.1 mg/m^3. To Mancuso, the data suggested that trivalent chromium may be at least partly responsible for the perforations. However, later studies have not supported this theory.

In a study reported by Lumio in 1953, 33 chromium platers were examined in workplaces where the highest measured concentration of chromium trioxide was 0.003 mg/m^3, and four were found to have septal perforation.[195] Of the 33 workers examined, only six had what the author considered normal noses. In view of the low chromium concentration, the author suggested that the lesions resulted from periodic high concentrations that occurred when ventilation of the tank failed.

Ulceration and atrophic rhinitis were described as occurring in anodizing operators exposed to concentrations of chromic acid mist ranging from 0.09 to 1.2 mg/m^3 (as CrO_3).[121,373]

A Russian article by Kuperman regarding maximal allowable hexavalent-chromium concentrations in atmospheric air reports on an experiment using 10 apparently normal persons exposed to hexavalent-chromium aerosol concentrations ranging from 0.0015 to 0.04 mg/m^3.[177] Short periods of inhaling air containing hexavalent chromium at 0.01–0.024 mg/m^3 sharply irritated the nose. The most sensitive person responded at a concentration of 0.0025–0.004

mg/m^3, as shown in Table 7-4. Whether this was a reaction to chromium or to the acidity of the aerosol is not known.

Vigliani and Zurlo reported septal perforation in workers exposed to chromates and chromic acid in concentrations of 0.11–0.15 mg/m^3.[354]

Hanslian *et al.* reported on otolaryngologic examinations of 77 persons exposed to chromic acid aerosol during chrome plating and found 19% to have septal perforation and 48% to have nasal mucosal irritation.[135] Papillomas of the oral cavity and larynx were found in 14 persons who averaged 6.6 years of exposure to an air chromium concentration of 0.4 mg/m^3. Histochemical examinations of the excised tumors confirmed the diagnosis of papilloma; however, there were no signs of atypical growth or malignant degeneration.

In summary, most studies are in agreement that hexavalent chromium in some concentrations produces ulceration and perforation of the nasal septum. The relation of this condition to allowable or safe concentrations is discussed in Chapter 11.

Respiratory Cancer

All the evidence available indicates that the only important long-term effect of hexavalent chromium is an increased incidence of lung cancer in the manufacture of dichromate from chromite ore. The risk has been established for the industry in Germany, the United States, and Great Britain, and available knowledge has been published.[16,17,32,33,36,97] The incidence of cancer at other sites in the body is not increased; nor does the risk appear in the user industries, although isolated cases have been reported.[152]

Epidemiologic studies of the relation between chromium and respiratory cancer are considerably more complex than studies of ulceration and perforation of the nasal septum. Respiratory cancer usually occurs only after some years of exposure (the average duration in Baetjer's[16] study was 17 years) and may not appear until long after the end of

TABLE 7-4 Threshold of Chromium Aerosol Irritation Effect[a]

	Concentration of Hexavalent Chromium, mg/m^3	
No. Observations	Minimal Perceptible	Maximal Nonperceptible
2	0.0040	0.0030
5	0.0030	0.0025
3	0.0025	0.0015

[a]Derived from Kuperman.[177]

exposure. Hence, the extent and type of exposure recorded at the time of diagnosis cannot be used as a measure of the exposure at the time of onset. Furthermore, because lung cancer has a multiplicity of causes, unlike ulceration and perforation of the nasal septum, a particular case cannot be attributed to chromium exposure.

The literature regarding the role of chromium in lung cancer up to 1950—mainly of German, British, and American origin—was reviewed by Baetjer.[16] Most of the following material was taken from her publication.

The first two cases of lung cancer in German chromate workers were observed in 1911 and 1912, but were not reported until 1932.[186] These men had been employed in an old chromate plant of the I. G. Farbenindustrie at Ludwigshafen, which was closed in 1923. It was stated that they had worked in the section of the plant in which dichromates were used for oxidizing anthracene and its derivatives into compounds with a quinone structure in the manufacture of alizarin dyes. The reduced chromium was regenerated into a dichromate, chromic acid, or chrome alum. In 1935, Pfeil reported that five additional cases had occurred in about 1930–1935 in workers of the same plant.[246] He did not state in which part of the plant the subjects had been employed, nor their occupation since 1923. Thus, there were a total of seven reported cases in this plant's workers.

Twenty cases of lung cancer were reported to have occurred between 1929 and 1938 in chromate workers of another old chemical plant, which was closed in 1931 (the I. G. Farbenindustrie at Griesheim).[4] Only 10 of these patients were employed in the chromate-producing section of the plant; one of the 10 had left the chromate industry and had worked in the aniline industry for 5 years preceding the diagnosis of cancer.[102] The remaining 10 men did not work directly in the chromate plant, but it was claimed that they had been exposed intermittently to chromates.

In addition to the 27 cases of lung cancer reported from these two old plants, several other cases were described in the German literature between 1936 and 1938. Teleky reported one case, but did not give the location of the plant.[329] Alwens and Jonas reported that one case had occurred in the plant of the I. G. Farbenindustrie at Uerdingen, one at Bitterfeld, and possibly one at Leverkusen.[5] However, the report of a British survey of industrial medicine in German factories in 1947[115] stated that there had been 10 workers with lung cancer in the chromate plant of the I. G. Farbenindustrie at Leverkusen, of whom seven had already died and three were ill at the time of the survey. The same report stated that "in the last 30–40 years there have been 14 cases of cancer of the lung" in the chromate plant at Uerdingen. No further description of these cases was given.

Therefore, at least 52 cases of cancer of the respiratory tract among workers in the chromate-producing plants of Germany were reported in the literature before 1950.

In 1943, eight cases of pulmonary carcinoma were described by Gross and Kölsch as having occurred in three chrome-pigment plants of Germany, and two cases in another similar plant were reported by Letterer, Neidhardt, and Klett in 1944.[124,187]

The only case of chromium-related lung cancer in the German literature that was not from either the chromate-producing or the chrome-pigment industry was described by Baader.[13] The patient had applied chrome pigment by hand or by spray for many years. The type of compound used was not mentioned.

No data had been published in the German literature by the time of Baetjer's study on the death rates associated with lung cancer in chromate workers or on the incidence of lung cancer in chromate workers, compared with other workers. Attempts had been made to obtain some figures, but these were of little value. Alwens reported that from 1929 to 1938 there had been 19 or 20 cases of lung cancer among the chromate workers of Griesheim.[4] In the 30-year period preceding this study, only three or four deaths from lung cancer occurred among men in the same locality (male population, 13,000) who were not employed in the chemical industry. These figures were obtained by a questionnaire sent to the doctors in the community. Gross and Kölsch stated that the German plants producing chromates since 1880 had employed about 2,000 men, 1,000 of whom had worked in the chromate industry for long periods.[124] They indicated that 39 cases of primary pulmonary carcinoma occurred among these 1,000 men.

In contrast with those reports, a few investigations in the German chromate industries did not reveal the presence of pulmonary cancer in the workers.[185] For example, although Lukanin performed physical examinations with roentgenologic and laboratory studies on 350 workers in a German chromate-producing plant in 1930, he did not mention bronchial carcinoma.[194] However, inasmuch as lung cancer is often fatal within a short time after becoming diagnosable, negative cross-sectional studies of small groups are inconclusive. Lehmann summarized the literature on the relation of chromium to health in 1913 and again in 1932, but did not mention any cases of pulmonary cancer.[185,186]

Cancers of other tissues in chromate workers were reported in only a few instances. Teleky expressed the opinion that there was a high rate of cancer of the digestive tract, as well as of the lung, in chromate workers; he found five such cases in men who had worked in a chromate plant for 2–26 years.[329] No other such cases appear to have been reported in the German literature, and no cases of cancer of the skin have been reported in the chromate workers, although chromium compounds often caused dermatitis and severe skin ulcers.

In general, the German literature suggested that some occupational factor may have been responsible for bronchogenic carcinoma in chromate workers, but does not provide any definite proof that the incidence of pulmonary carcinoma was abnormally high in the chromate-producing industry or that chromate compounds alone were the responsible substances in the cases seen. Men in whom lung cancer developed were also exposed to other substances, some of which may have been carcinogenic, such as anthracene and its derivatives and the quinone compounds. Moreover, it is probable that some of these men developed lung cancer for reasons not associated with their occupation.

In Britain in 1951, Bidstrup reported on chest x-ray examinations of 724 workers in the British chromate-producing industry and found one pulmonary carcinoma.[34] Comparison with findings in mass chest x-ray surveys carried out in Great Britain in 1949 suggested some excess in the prevalence of pulmonary carcinoma in workmen in the chromate industry. In 1956, Bidstrup and Case reported on a 6-year follow-up for mortality among workers from three chromate-producing plants in Great Britain; the results are shown in Table 7-5.[36] Twelve deaths from lung cancer were noted, compared with 3.3 expected on the basis of experience with a comparable section of the total population. Deaths from neoplasms elsewhere in the body and deaths from other causes were not found to be significantly excessive.

The first study of lung cancer in chromate workers in the United States was made in 1948 by Machle and Gregorius at the request of

TABLE 7-5 Observed and Expected Deaths among 723 Workers in Three Chromate Plants, by Cause of Death[a]

	No. Deaths	
Cause of Death	Observed	Expected
Lung cancer	12	3.3
Cancer at other sites	9	7.9
Other causes	38	36.3
TOTAL (all causes)	59	47.5

[a]Derived from Bidstrup and Case.[36]

the chromate-producing industry.[197] Mortality data were obtained from the records on group life-insurance policies of chromate companies and were compared with similar data from industrial policyholders of the Metropolitan Life Insurance Company. Death from cancer of the lung and bronchi was reported to have occurred between 1937 and 1947 in 42 employees of the five chromate-producing plants in the United States. There were also a few cases of cancer of the upper respiratory tract or the oral region among older workers, but the rates were not conclusive. The results of the comparison with other industrial policyholders are shown in Table 7-6. Of all deaths among employees of the chromate-producing plants, 21.8% were reported as due to cancer of the respiratory system, compared with 1.4% of deaths among the controls. Machle and Gregorius also compared lung-cancer rates among chromate workers with rates among workers in an oil-refining company. The results are shown in Table 7-7. Mortality from lung cancer among chromate workers was about 40 times greater than that among oil-refinery workers at ages 50 or less and about 20 times greater at ages over 50. Three of the five plants studied exhibited high death rates for cancers of the oral region, nose, and pharynx, but the authors felt that the rates were not conclusive.

In 1950, Baetjer reported a retrospective study of pulmonary carcinoma in chromate workers.[17] This study was based on the records of two Baltimore hospitals near a chromate-producing plant. The number of chromate workers among patients with lung cancer, as confirmed by autopsy and biopsy, in these two hospitals during the preceding 15 and 20 years (1926–1946) was compared with the number of chromate workers among other hospitalized groups selected as controls. The results are shown in Table 7-8. Statistical analysis indicated that the percentage of chromate workers was significantly higher in the lung-cancer group than in other hospitalized groups, used as controls, or than would

TABLE 7-6 Deaths Due to Cancer and Cancer of the Lung among Chromate Workers and Controls[a]

	Chromate Workers		Controls[b]	
Cause of Death	No. Deaths	% of All Deaths	No. Deaths	% of All Deaths
All cancers	66	34.2	115	15.7
Cancers of the lung	42	21.8	10	1.4
TOTAL (all causes)	193	100.0	733	100.0

[a]Derived from Machle and Gregorius.[197]
[b]Metropolitan Life Insurance Company industrial policyholders, 1946.

TABLE 7-7 Annual Mortality Rates for Lung Cancer[a]

	Annual Mortality Rate per 100,000 Males	
Type of Worker	Age ≤ 50 Years	Age > 50 Years
Chromate workers	197	480
Oil-refinery workers	5	22

[a]Derived from Machle and Gregorius.[197]

be expected on the basis of the number of men employed at the chromate-producing plant.

A prospective study of deaths from pulmonary carcinoma in chromate workers similar to that of Machle and Gregorius was made by Mancuso in 1949.[203] He examined the records of all employees who had worked for a year or more between 1931 and 1949 in a chromate-producing plant in Ohio, to determine how many had died and the causes of death, as listed on death certificates. Of 33 chromate workers in this group who had died, six (18.2%) had had a diagnosis of pulmonary carcinoma. In a control group consisting of all males who had died between 1937 and 1947 in the county with the chromate plant, only 1.2% of all deaths were due to cancer of the respiratory system.

Mancuso (personal communication), in a follow-up of the original prospective study of employees of a small chromate plant, so far has found among 160 deaths the following distribution of causes of death: 38, cancer of the respiratory system; two, cancer of the nasal sinuses; and one each, cancer of the nasopharynx, cancer of the kidney, and

TABLE 7-8 Association of Occupational Chromate Exposure with Lung Cancer and with Other Causes of Hospitalization[a]

		Patients with Chromate Exposure	
Patient Group	Total No. Patients	No.	%
Johns Hopkins Hospital:			
Lung cancer	198	7	3.53
Other causes of hospitalization (random sample)	226	0	0
Baltimore City hospitals:			
Lung cancer	92	4	4.35
Other causes of hospitalization (random sample)	499	0	0

[a]Derived from Baetjer.[17]

TABLE 7-9 Deaths from Respiratory Cancer among Male Members of Sick-Benefits Organizations in Six Chromate-Producing Plants, by Age[a]

	No. Deaths	
Age, yr	Observed	Expected
15–44	4	0.1
45–54	12	0.4
55–74	10	0.5
TOTAL	26	0.9

[a]Derived from Federal Security Agency.[97] Statistics exclude deaths due to cancer of the larynx.

multiple myeloma. However, no data on the control group are available for comparison.

The United States Public Health Service has reported on the morbidity and mortality experience of male members of sick-benefits associations in several chromate-producing plants.[97] Over the 9-year period, 1940–1948, there were nearly 29 times as many deaths from respiratory cancer as would have been expected on the basis of the mortality experience of all males in the United States, as shown in Table 7-9. Cancer at other sites was not excessive.

The Public Health Service study also included a report on the medical examinations of 897 chromate workers. Ten persons, with a mean exposure of 22.8 years, were considered to have bronchogenic carcinoma. Because some had had positive diagnoses before the survey, this cannot be compared with surveys of apparently healthy populations. Nevertheless, the finding of so many cases in this relatively small population represents an extraordinarily high incidence.

In a fairly recent study, Taylor followed a group of chromate workers of three chromate plants over a 24-year period using Old Age and Survivors Disability Insurance records from the United States Social Security Administration.[328] The results are shown in Table 7-10. For this cohort, the observed : expected death ratio for lung cancer (base, total male experience in the United States) was 8.5 : 1. Excess mortality for all causes of death was due mainly to lung cancer, although slight excesses were also noted for other causes of death—a ratio of 1.3 : 1 for other cancer and a ratio of 2.4 : 1 for other respiratory diseases.

Although no dose–response relation between chromium and lung cancer is known, it seems certain that one exists. Not only are exposed workers at a greatly increased risk of developing lung cancer, but the risk increases with duration of exposure. In Taylor's study, there was a definite relation between the death rate for respiratory cancer and cumulative years of experience in the chromate industry (with age held constant). For a group of men with a mean age of 52.5 years, for example, the rate of death from respiratory cancer increased from 106.2 per 100,000 for men with 5–9 years of experience to 400 per 100,000 for men with 20–24 years of experience.

The carcinogen (or carcinogens) present in the manufacturing process has not been identified. The composition of the chromium materials produced in the old chromium-chemical manufacturing plants differed somewhat, but the processes were generally the same. Chromate ore was the basic material. After roasting with soda ash, with or without lime, the desired end product was sodium chromate, from which dichromates and other chemicals were produced. Considerable residue from the roast remained after the sodium chromate was leached out by water, which was reused in most plants. The solubility and the valence of the principal materials are indicated in Tables 7-11 and 7-12. Circumstantial evidence based on study of the work records of individual patients led Bidstrup to believe that dust arising from furnaced roast and the residue after monochromate is removed are the most likely sources of the carcinogen. On the basis of their initial animal experiments (see Chapter 8), Laskin, Kuschner, and Drew suggested the compounds listed in Table 7-13 as the chromium compounds most promising for investigation of carcinogenic properties.[181]

TABLE 7-10 Observed and Expected Deaths among Chromate Workers Aged 14–64 Years, by Cause of Death[a]

	No. Deaths	
Cause of Death[b]	Observed	Expected
Tuberculosis (001–019)	9	8.0
All cancer (140–205)	103	32.2
Respiratory cancer (160–165)	71	8.3
All other cancer	32	23.8
Cardiovascular diseases (330–337, 400–468)	89	99.4
Other respiratory diseases (470–527)	19	7.8
All other causes	38	61.5
TOTAL (all causes)	258	209.1

[a]Derived from Taylor.[328] Subjects born 1890 or later.
[b]Numbers refer to *Manual of the International Statistical Classification of Diseases, Injuries, and Causes of Death* (Seventh Revision).[369]

TABLE 7-11 Solubility of Chromium in Various Plants, by Parent Material[a]

		Chromium, % of total sample		
Parent Material	No. Samples	Water-Soluble	Acid-Soluble and Water-Insoluble	Acid-Insoluble
Ore	6	0.01– 0.07	0.01–0.04	29.6 –31.6
Mix	4	0.02– 0.52	0.02–0.13	4.4 –15.9
Secondary mix	1	0.88	2.66	2.5
Roast	6	1.11– 8.63	0 –2.12	2.02–10.41
Secondary roast	2	4.14–12.3	0.79–1.1	0.16
Residue	3	0.4 - 0.75	0 –2.7	2.4 - 9.15
Secondary residue	1	2.23	0.49	0

[a]Derived from Federal Security Agency.[97]

Gross[123] and Machle and Gregorius[197] considered monochromates to be the chromium compounds responsible for lung cancer. Bauer thought the responsible substance was the free chromic acid or the alkali salts, especially the dichromates.[23] Kölsch believed that the monochromates and dichromates were the responsible factors in pulmonary cancers and that the chromites were not harmful, because they are scarcely soluble

TABLE 7-12 Hexavalent Chromium in Samples of Ore, Roast, and Residue Materials by Solubility[a]

Solubility of Sample Type	Total Chromium, % of sample	Hexavalent Chromium, % of sample
Ore		
Water-soluble	0.02	0
Acid-soluble and water-insoluble	0.02	0
Acid-insoluble	31.30	0
TOTAL	31.34	0
Roast		
Water-soluble	9.10	9.10
Acid-soluble and water-insoluble	2.80	0.03
Acid-insoluble	2.20	0
TOTAL	14.10	9.13
Residue		
Water-soluble	0.40	0.40
Acid-soluble and water-insoluble	2.70	0.10
Acid-insoluble	2.40	0
TOTAL	5.50	0.50

[a]Derived from Federal Security Agency.[97]

in the body.[168] Among pigment workers, Gross and Kölsch believed zinc chromate was the cause of lung cancer, because it was more soluble than barium or lead chromate.[124]

Mancuso and Hueper gave serious consideration to chromite ore as a potential carcinogenic agent, as well as chromium pigments and chromium alloys.[204] They believed that the insoluble chromium compounds are retained in the lung for long periods and may give rise there to the production of pneumoconiotic changes. Mancuso and Hueper have noted a lack of good epidemiologic studies on workers exposed only to trivalent chromium compounds.[204]

The United States Public Health Service study deals extensively with the form of chromium that might produce lung cancer.[97] A refractory plant using chromite ore to make chromite brick was studied, and deaths were observed over a 14-year period, 1937–1950. Only one case of lung cancer was observed—exactly what was expected if lung-cancer death rates for the United States as a whole were applied to the plant population. It was concluded that chromite was not carcinogenic, despite the earlier suggestion by Mancuso and Hueper[204] that chromite ore was the potential carcinogenic agent. The Public Health Service report also casts doubt on the role of soluble hexavalent chromium because of its extreme solubility and consequent rapid dissipation. Other derivatives of chromite ore were strongly suspect—specifically compounds that are acid-soluble and water-insoluble.

In 1972, a study was made of approximately 300 workers who had been exposed for the preceding 20–25 years in a German plant to trivalent chromium (as chromic oxide and chromic sulfate). Clinical, x-ray,

TABLE 7-13 Valence and Solubility of Selected Chromium Compounds[a]

Material	Formula	Valence	Solubility
Process residue:			
Chromates	CrO_4^{-2}	6	Water-soluble
Chromate–chromite complex	$CrO_4^{-2}\ Cr_2O_4^{-2}$	3,6	Water-insoluble and acid-soluble
Chromic oxide	Cr_2O_3	6	Acid-insoluble
Calcium chromate	$CaCrO_4$	6	Moderately water-soluble and alcohol-soluble
Chromic chromate	$Cr_2(CrO_4)_3$ $x(Cr_2O_3)y(CrO_3)$	3,6	Water-soluble;[b] forms colloids
Chromic oxide	$(x)Cr_2O_3$	3	Water-insoluble and acid-insoluble
Chromic trioxide	$(y)CrO_3$	6	Water-, alcohol-, ether-, and acid-soluble

[a]Derived from Laskin *et al.*[181]
[b]CrO_3 added to freshly prepared $Cr(OH)_2$.

and blood studies, disability rates, and other health indices were compared with corresponding findings in a group of control workers not exposed to chromium. Only preliminary results are available. No significant differences were detected between the chromium-exposed group and the control group in the rate of respiratory illness, x-ray findings, clinical picture, or blood studies. No cases of lung cancer were known to have occurred in the plant during a 24-year period preceding the study (U. Korallus, personal communication).

Because no investigation of cancer in the chromium-using industries (such as plating and anodizing) has been published, it cannot be stated that exposures in these industries do not increase the risk of lung cancer. However, in view of the widespread knowledge of the relation of chromium to cancer and the many industrial uses of chromium chemicals, the lack of reported cases strongly suggests that the chromium-related lung-cancer problem is limited to the chromate-producing industry.

The symptoms, signs, clinical course, x-ray appearance, methods of diagnosis, and prognosis of lung cancer associated with chromium are similar to those of lung cancer due to other causes. As with all cancers, there is a latent period between first exposure and the onset of disease. Bidstrup and Case[36] found the average latent period to be 21 years, the period between first exposure and onset being less than 10 years in two cases and more than 30 years in two cases; Baetjer[16] recorded one case in which only 4 years elapsed between first exposure and onset, and another in which the interval was 47 years.

The cancers found in chromate workers are not of any single histologic type. In a worldwide study of 123 lung cancers in chromate workers, Hueper[152] found 46 squamous-cell carcinomas, 66 round-cell carcinomas, and 11 adenocarcinomas.

Among cases observed in Great Britain, tumors tended to arise peripherally, and squamous-cell carcinomas predominated. Faulds (quoted by Bidstrup[33]) examined the lungs of 28 British chromate workers, 11 of whom died from lung cancer. Squamous-cell carcinoma was found in nine, and adenocarcinoma in two. Fibrosis of lung tissue was present in two of these cases, and squamous metaplasia in two others. In none of the lung-cancer cases was either squamous metaplasia or interstitial fibrosis present at sites in the lung other than those involved in the neoplastic change.

Of 10 cases that came to autopsy in the United States, the tumors appeared to arise chiefly from the main bronchi and were described histologically as oat, squamous, undifferentiated epithelial, or anaplastic tumors. In one case, two distinct tumors were found—a squamous-cell tumor of the bronchus and an oat-cell tumor of the pleura.[16]

When the tumor arises peripherally, the prospect of successful treatment by operation–lobectomy or pneumonectomy–is enhanced, provided that time is not lost in attempts to confirm diagnosis before surgery. In Great Britain, two patients operated on in the mid-1950's are surviving and in full-time work, as are several similarly treated in 1960–1962. Others are well and working 5 years after operation (Bidstrup, unpublished data).

Of the methods of early diagnosis available, routine x ray remains the most reliable. When the interval between x rays has been 8 months, most cases have been revealed at the symptom-free stage. However, in some workers, advanced cancers have appeared without previous evidence, even when there have been such periodic roentgenologic examinations. All persons at risk must be under competent medical supervision, so that they can be referred for investigation, including x ray, whenever episodes of illness occur for which no obvious cause can be found. In the past, sputum cytology has not proved helpful for early diagnosis; more sophisticated methods and techniques are being developed for collection of specimens and identification of cells, which may make this method become more valuable. Most authorities believe that urine, blood, and tissue analyses for chromium are not of diagnostic value, except possibly to confirm exposure. Mancuso[202] reported that ranges of chromium concentrations in the blood and urine of chromate workers were 0.5–17 μg/100 g and 0–380 μg/liter, respectively.

Other Respiratory Effects

According to Bidstrup (personal communication), dust or mist containing hexavalent chromium irritates mucous membranes, causing sneezing, rhinorrhea, irritation and redness of the throat, and generalized bronchospasm. Sensitization may develop, resulting in typical asthmatic attacks, which recur on later exposure even when exposure is to much lower concentrations.

Exposure to higher concentrations may give rise to more serious symptoms, including cough, headache, dyspnea, and substernal pain. Generalized rhonchi and moist rales may be heard on examination of the lungs. Experience suggests that bronchospasm occurring in a man working with chromates is likely to be due to chemical irritation of the air passages, and not to "bronchitis," although it is commonly so diagnosed. It is not known whether repeated episodes of chemical irritation will lead ultimately to chronic bronchitis, but the probability is that it will do so.

Some German investigators reported that there were no marked clinical symptoms in persons exposed to chromate dust.[102,185] Other German workers have reported that the prolonged inhalation of chromate dust caused chronic irritation of the respiratory tract and resulted in such manifestations as congestion and hyperemia, chronic catarrh, congestion of the larynx, polyps of the upper respiratory tract, chronic inflammation of the lungs, emphysema, tracheitis, chronic bronchitis, chronic pharyngitis, and bronchopneumonia.[5,103,168,186,202] They described the roentgenograms as showing enlargement of the hilar region (often on only one side), enlargement of the lymph nodes, increase in peribronchial and perivascular lung markings, and adhesions of the diaphragm. Some German doctors claimed that a characteristic pneumonoconiosis resulted from exposure to some chromates.[187,194]

However, medical officers of the United States Public Health Service who examined 897 workers in the chromate-producing plants reported that, although the workers had a higher incidence of severely red throats and a higher incidence of pneumonia, they did not show any increase in the incidence of other respiratory diseases, compared with control groups.[97] Furthermore, there was no evidence of excessive pulmonary fibrosis in these chromate workers, although bilateral hilar enlargement was noted. Whether the various lung changes described as occurring in chromate workers represent merely a nonspecific reaction to irritating material or a specific reaction to chromium compounds cannot be stated at present. Many of these conditions occur widely in the general adult population, especially among older people.

Two cases of acute pulmonary complications involving the deeper pulmonary structures after inhalation of massive amounts of chromic acid mist have been described.[219] Whether they were due to the chromium or to the acid is not known. The estimated chromic acid concentration in the mist was 20–30 mg/m^3. The symptoms included cough, chest pain, some dyspnea, pleural effusion, and loss of weight. In another study, atrophic rhinitis was reported in 5–10% of a group of workmen exposed to the mist of a 5% chromic acid solution.[373] Hyperemia, swelling, congestion, and nasal catarrh occurred in some of these persons.

Various other conditions have been attributed to chromium, but, in most cases, the etiologic relation to chromium is doubtful because of the presence of other chemicals. In Norway, asthma was reported as a complaint among workers employed for more than 10 years in a ferrochromium plant.[45] Four cases of pulmonary disease with nodular fibrosis and ventilatory impairment were described among workers in one chromium-alloy plant,[257] but no such cases were found in another, similar plant.[250] Mancuso[202] reported that chromate workers frequently showed excessive liability to inflammatory and ulcerative conditions of the gastrointestinal tract caused by ingestion of chromates.

Pulmonary markings (ground-glass types 1 and 2) were reported in workers in a chromite mine and concentration plant and were attributed to chromite dust. However, free silica dust also was present in the air. No clinical or roentgenologic evidence of fibrosis was found in the chromate chemical workers.[97]

SKIN EFFECTS

Writing in the *Edinburgh Medical and Surgical Journal* in 1827, William Cumin related that he had been consulted 5 years earlier by two dyers whose wrists and arms had developed deep ulcers after immersion in a solution of "bichromate of potass."[70] A few years later (1833), Ducatel[85] described skin ulcers of the hands and arms of workmen employed in a chromate-manufacturing plant in the United States (Baltimore). In France, Bécourt and Chevallier[24] in 1863 and Delpech and Hillairet[79] in 1869 reported on the cutaneous and mucous-membrane effects observed among dichromate workers.

Early in this century, there began a steady flow of information dealing with the health effects of chromium compounds, with contributions from Hermanni,[140] Legge,[184] Fischer,[102] Lehmann,[185] DaCosta *et al.*,[75] and many others. Most of the early reports stressed the corrosive or ulcerative action of the chromates, no doubt because the ulcers were readily recognizable. They appeared in a high percentage of workmen handling chromates, so a cause-and-effect relation was evident.

With the expanding use of chromates in industrial processes, more workmen became exposed to them. Consequently, more cases of cutaneous effects occurred, and skin changes in addition to the well-known ulcers were observed and reported in the United States,[239,317] Germany,[94] and Great Britain.[363]

It gradually became apparent that workmen exposed to chromates in their manufacture, in textile dyes and paint pigments, in photolithography and plate printing, in leather tanning, in metal and wood polishing, in metal plating, in blueprinting, in cement masonry, and in other pursuits were prone to incur various types of skin injury from one or more of the chromate compounds. Thus, since the early report by Cumin, there has developed a voluminous world literature, first emphasizing the ulcerative or destructive lesions associated with chromate action on skin, then the occurrence of allergic responses of the skin to chromates, and more recently the possible pathogenetic mechanism involved in the chromium-allergy phenomenon.

It is generally agreed that chromium reactions of skin can be classified in the following dermatologic categories: corrosive reactions, including ulcers and stigmata (scars), and sensitization reactions, including eczematous contact dermatitis (allergic).[277] A suggested modification of the classification would categorize the reactions as follows: primary irritations, including ulcers (corrosive reactions), scars, and nonulcerative contact dermatitis; and allergic contact dermatitis, including both

TABLE 7-14 Potential Occupational Exposures to Chromium[a]

Abrasive makers	Jewelers
Acetylene purifiers	Laboratory workers
Adhesive workers	Leather finishers
Airplane sprayers	Linoleum workers
Alizarin makers	Lithographers
Alloy makers	Magnesium treaters
Aluminum anodizers	Match makers
Anodizers	Metal cleaners
Battery makers	Metal workers
Biologists	Milk preservers
Blueprint makers	Oil drillers
Boiler scalers	Oil purifiers
Candle makers	Painters
Cement workers	Palm oil bleachers
Ceramic workers	Paper waterproofers
Chemical workers	Pencil makers
Chromate workers	Perfume makers
Chromium-alloy workers	Photoengravers
Chromium-alum workers	Photographers
Chromium platers	Platinum polishers
Copper etchers	Porcelain decorators
Copper-plate strippers	Pottery frosters
Corrosion-inhibitor workers	Pottery glazers
Crayon makers	Printers
Diesel locomotive repairmen	Railroad engineers
Drug makers	Refractory-brick makers
Dye makers	Rubber makers
Dyers	Shingle makers
Electroplaters	Silk-screen makers
Enamel workers	Smokeless-powder makers
Explosive makers	Soap makers
Fat purifiers	Sponge bleachers
Fireworks makers	Steel workers
Flypaper makers	Tanners
Furniture polishers	Textile workers
Fur processors	Wallpaper printers
Glass-fiber makers	Wax workers
Glass frosters	Welders
Glass makers	Wood-preservative workers
Glue makers	Wood stainers
Histology technicians	

[a]Derived from Milby *et al.*[222]

eczematous and noneczematous. All the cutaneous changes listed can be observed in the working population exposed to chromates. As to chromium dermatoses among the general population, allergic contact dermatitis is the form of cutaneous change usually seen, whereas chromium ulcers and nonulcerative contact dermatitis rarely occur.[359]

We do not know the exact incidence of dermatoses due to chromium in or out of industry, but the present industrial use of chromium is widespread. The scope of potential occupational exposure to chromium is indicated by Table 7-14. Despite this awesome exposure potential, the frequency of dermatoses due to chromates today seems to be considerably lower than it was before and during World War II. There are no accurate statistics on the subject, but it would be readily apparent to anyone who had visited large and small industrial plants over the last 25 years that hygienic controls have greatly improved in many operations, including those in which chromates are handled.[37]

Primary Irritant Dermatoses

Chromium ulcers are commonly called "chrome holes" or "chrome sores"; DaCosta *et al.*[75] referred to them as "acid bites." These rather characteristic lesions result from contact with chromic acid, sodium or potassium chromate or dichromate, or ammonium dichromate. An ulcer develops if the chromium compound, as either a dust or a liquid, is deposited or splashed on any break in the skin—e.g., an abrasion, a scratch, a puncture, or a laceration.[302] Ulceration is more likely to occur among workers who encounter strong concentrations of chromic acid, sodium or potassium chromate or dichromate, or ammonium dichromate.[302] Favored sites for ulcer development are the periungual (nailroot) areas, the creases over the knuckles, finger webs (particularly between thumb and forefinger), the backs of the hands, and the forearms.[24,75,79,85,102,140,184,185,277,363] Of course, an ulcer can appear anywhere on the skin if the conditions are appropriate—i.e., traumatized skin and sufficient exposure to chromic acid or one of the chromates. Edmundson[89] reported the dermatologic findings in one of the plants of the Public Health Service study conducted in six chromate-manufacturing plants in Maryland, New Jersey, Ohio, and New York: Of 285 workmen examined, 198 (60.5%) had chrome ulcers or scars, and 175 (61.4%) had nasal septal perforations. Chrome ulcers and their scars were distributed as follows: hands, 46.3%; arms, 19.5%; ankles and feet, 10.1%; legs, 6.0%; back, 4.1%; knees, 3.6%; thighs, 3.3%; abdomen, 3.0%; face, 1.9%; neck, 1.6%; and chest, 0.6%.

The full report of the study[97] provided a concise and accurate description of these lesions:

> This type of lesion occurs more readily if there is a break in the continuity of the skin, such as an abrasion, scratch or a laceration.[85,111,208,336] * The break in the skin becomes exposed to the chrome compound and the typical lesion ensues. The lesion has been described as being a round, nonspreading, deeply penetrating ulcer which has a hard, well-defined circular and raised border. The central crater is clean cut and leads downward to a base covered with exudate or tenacious crust. Once a chrome sore has developed it is slow to heal, and if exposure continues, it may persist for months. The healing process invariably leads to scar formation which is flat and atrophic. The lesion begins as a painless papule of pinhead size that gradually enlarges to form the mature lesion which may vary from 3–10 mm in diameter and may extend to considerable depth.[158]

DaCosta *et al.*[75] described severe ulcerative changes that resulted in joint penetration, which in some instances required amputation to stop the destruction. Ordinarily, the duration of a chrome sore, if not deep, is about 3 weeks, provided that exposure is discontinued.[363] There is no evidence that the ulcers undergo malignant degeneration. Furthermore, the presence of chromium ulcers has no influence on the development of allergic sensitization to chromates.[89]

In the past, ulcers were reported in connection with various exposure sources. For example, in England in 1926, White[363] stated that 55 chromium ulcers had been attributed to various sources, as follows: manufacture, 2; dyeing and finishing, 33; tanning, 2; French polishing, 7; manufacture of chromium dyes, 5; and other industries, 6. The same frequency of ulcers in most of the same trades would be improbable today, because work processes have changed and hygienic controls are better. A few years ago, tannery workers were at greater risk than today, because they were heavily exposed to potassium dichromate in the two-bath tanning process. Now, most chromium tanning is done by a one-bath method that uses trivalent basic chromium sulfate, which is nonulcerogenic.[279] Exposures to chromium compounds with ulcerogenic potential today are confined largely to chromate manufacture and to electroplating operations.[302]

Therapy of chromium ulcers, whether cutaneous or nasal, has not been highly successful. Any number of antidotal remedies have been devised and tested. DaCosta *et al.*[75] used silver nitrate to precipitate the chromium and form silver chromate. Blair[38] proposed washing the hands with a 5% solution of sodium hyposulfite to neutralize the chro-

*Reference numbers in this extract have been changed to conform with this volume.

mium. Maloof[201] claimed success in treating chromium ulcers by using an ointment containing ethylenediaminetetraacetic acid (EDTA). Bidstrup (personal communication) states that a striking reduction in the incidence of chromium ulcers of the skin has been recorded for the manufacturing industry in Great Britain since 1956, when an ointment containing 10% sodium calcium edetate (EDTA, Versenate) was adopted for prophylaxis and therapy. All breaks in the skin, however slight, are cleaned, the ointment applied, and the area covered with an impervious dressing, which is replaced as necessary until healing is complete. Ulcers occur seldom; when they do occur, they are small and heal rapidly. Before 1956, much time was lost from work because of chromium ulceration of the skin; lost time is exceptional now and occurs only when damaged skin areas have not been treated promptly and adequately. It is difficult to persuade employees to apply Versenate ointment intranasally before starting work. Nasal ulceration therefore occurs more frequently; but if it is treated, healing will usually take place without perforation. Samitz *et al.*,[280] working in chromium research for many years, first proposed the use of a mixture of sodium pyrosulfite, tartaric acid, glucose, and ammonium chloride as an "antichrome" agent. It was intended as a preventive measure for chromate dermatitis, but it had several limitations.

More recently, Samitz *et al.*[283,285-287] have worked with ascorbic acid as an antichromium agent. They believe its efficiency as a preventive measure lies in its ability to reduce hexavalent chromium to the trivalent form, which it then complexes. Pirozzi *et al.*[253] have used a 10% aqueous solution of ascorbic acid, which accelerated the healing of experimentally induced chromate ulcers in guinea pigs. Protection was enhanced if the abraded skin was treated with ascorbic acid soon after it had been exposed to the hexavalent chromium solution. Even if there was a 30-min delay between chromium contact and the application of ascorbic acid, some benefit was provided. To act as an effective antichromium agent, the ascorbic acid must be freshly prepared every 3 or 4 days. More will be known about the efficacy of this substance after sufficient field experience has been gained.

Allergic Eczematous Contact Dermatitis

Chromic acid and the chromates are powerful skin irritants, and, in lower concentrations, the chromates are sensitizers. Acute primary irritant contact dermatitis occurs when workmen are exposed to the steam of boiling dichromate solutions.[302,363] White[363] described a diffuse erythematous dermatosis that resulted from dichromate, in which severe cases progressed to an exudative phase. He stressed that this form was different from and less common than the chronic dermatosis due to chromium. Bidstrup[35] states that workers in chromate-producing plants sometimes develop skin irritation—particularly at points of contact, such as neck and wrists—soon after starting work with chromium. Most of these cases clear quickly and do not recur, even when the workers return to work. In the Public Health Service study,[97] only a small percentage of the workmen examined had contact dermatitis.

Cases of allergic eczematous contact dermatitis due to chromate exposure probably occurred for many years before they were recognized for what they were. It is difficult to ascertain when the phenomenon was first recognized. In 1925, Parkhurst[239] reported a case of diffuse eczematous contact dermatitis in a woman working with potassium dichromate in blueprint production. She was skin-tested with a 0.5% solution of potassium dichromate on one thigh; 12 hr later, she developed a follicular erythematopapular rash at the test site. A similar test was applied to the other thigh, but the stain of the chromate was removed by sodium bisulfite about 30 min later; there was no reaction at this site. The patient returned to work and continued without dermatitis as long as she washed her hands in bisulfite solution every half-hour. Parkhurst concluded that potassium dichromate could affect the skin of susceptible persons and that the dermatitis can be averted by using sodium bisulfite wash. Smith[317] observed a case of chromium poisoning with manifestations of sensitization in a man who worked with ammonium dichromate. The subject reacted to a 1% ammonium dichromate patch test. He was later tested intradermally with a 0.5% solution of ammonium dichromate and within 6 hr developed a generalized severe cutaneous and systemic reaction. On the face of the evidence, it was unclear whether the reaction was caused by intoxication or an allergic sensitization to chromium.

Conclusive evidence of chromate sensitivity was elucidated in the work of Englehardt and Mayer,[94] who investigated chromium eczema in the graphic industry (lithography) by studying the problem in eight public and six private printing shops. Of 114 workers studied, 30 had eczema; of these, 84% reacted to a patch test with 0.5% potassium dichromate. None of these workers reacted to a trivalent test material. In 1944, Hall[128] called attention to the high frequency of contact dermatitis due to zinc chromate primer used by aircraft workers on aluminum and magnesium airplane parts that had been anodized (to prevent rust). A high percentage of these workers reacted to patch tests with potassium dichromate. Pirilä and Kilpiö[252] reviewed 45 cases of dermatitis caused by dichromates among woodworkers, cement and lime workers, radio-factory employees handling photostats, metal-factory workers, painters, polishers, furriers, and others. They observed that sensitization usually required about 6–9 months, but could occur in less than 3 months. In their cases, the hands were the sites commonly affected. They used 0.5% potassium dichromate for patch-test verification. (At that time, it was not known that cement contained chromium; the positive reactions to potassium dichromate among their cement workers were not interpreted in that light.)

In 1950, Jaeger and Pelloni[160] patch-tested 32 cement-eczema patients with potassium dichromate; 30 reacted. Of 168 patients with eczema from other causes, only 5% reacted to skin tests with the 0.5% potassium dichromate. They also analyzed several samples of cement and found minute quantities of chromium. They concluded that cement eczema was caused by chromium salts. Engebrigtsen[92] in 1952 published his investigation of the relation of hypersensitivity to dichromate among cement workers and concluded that cement eczema has a toxic, as well as an allergic, factor in its genesis. He also believed that the allergen was the water-soluble potassium dichromate found in most cements. The content of chromium in American cement was investigated by Denton *et al.*[80] (1954) and shown to parallel the chromium content of cement from other countries. Calnan[52] reviewed the cement-dermatitis problem in 1960. He demonstrated that soluble chromate was identifiable in almost all British cements and suggested that the chromate had become a part of the cement during its processing—e.g., from the chrome steel used for milling, from the refractory bricks in the kiln, or perhaps from the coal ash. He concluded that contact dermatitis due to cement was a primary irritant effect complicated by a supervening contact sensitivity to hexavalent chromium. Detection of chromium in cement and the patch-test reactions observed in cement dermatitis led to hundreds of clinical papers dealing with the subject in various countries.

Additional interest in chromate dermatitis came about as a result of the use of sodium and potassium dichromate in diesel locomotive radiators and cooling systems to prevent corrosion. Winston and Walsh[364] were first to investigate the problem and show the relation between the dichromate in the coolant and the dermatitis. It became clear that diesel machinists working in the locomotive repair shops came into contact with the radiator fluid or with powdered chromium sublimate from the leakage around gaskets, thereby becoming sensitized to the chromate antioxidant. Many cases occurred before the situation could be brought under control.

Periodically, reports of chromate dermatitis from unusual sources appear in the literature. One of these occurred in 1955, when Morris[223] first described dermatitis from a glue made from chromium trimmings and shavings. Such glue can cause dermatitis in those making or using it.

A few years later, Fregert[106] showed that chromate in matchheads could cause allergic eczematous contact dermatitis, particularly because the matchheads would partly dissolve when held by moist fingers. He later demonstrated[105] that book matches could contaminate the user's hands with chromate, especially in hot, humid weather; the matches would soften and flake, and the debris would collect in the user's pockets.

Lithographer's dermatitis was studied in the United States in 1959 by Levin *et al.*,[188] who showed after extensive patch testing that the most important causal substances were chromium compounds. Trauma and contact with soaps, solvents, and acids render the skin more susceptible to the action of irritant and allergenic chemicals encountered in this trade.

Engle and Calnan,[93] investigating an outbreak of dermatitis in an automobile factory in England, found that the 60 people involved were engaged in wet-sanding the primer paint applied to car bodies. They later ascertained, after appropriate patch testing, that zinc chromate was the offending agent. The dermatitis generally varied in appearance from simple erythema to dyshidrotic and exudative eczema; some cases showed nothing more than dry, scaling dermatitis. At about the same time in England, Newhouse[232] reported a survey of 230 automobile assemblers, 36% of whom reacted to one or more of the nine substances used in a battery of patch tests. Chromate sensitivity was four times more prevalent among the assemblers than among other employees. The source of the dermatitis was deduced as having been the chromate dip used as an antirust agent on nuts, bolts, screws, and washers.

An unexpected source of chromate contact dermatitis was reported by Fregert and Ovrum[107] in 1963 and by Shelley[305] in 1964. In both instances, chromium fumes generated from welding rods caused active contact dermatitis. In the Shelley study, the inhalation of the welding fumes exacerbated the dermatitis on the palms of a man who reacted to a patch test with potassium dichromate. It was later found that some welding rods contained up to 18% chromium.

Historically, allergic contact dermatitis due to chromium was unequivocally associated with compounds of hexavalent chromium, notably potassium and sodium dichromate and chromate and ammonium dichromate. In 1958, Morris[224] reviewed chromium dermatitis with particular reference to shoe leather and cited his experiences with several patients who demonstrated chromium allergy. Working on the basis that shoe leather was tanned by trivalent basic chromic sulfate, he demonstrated that four patients with shoe-leather dermatitis reacted to patch tests with 0.2% trivalent basic chromic sulfate. Two leather workers with a history of chromium dermatitis also reacted to the trivalent test material, whereas a diesel engine inspector exposed to hexavalent chromium reacted to the hexavalent, but not to the trivalent, chromium. Morris recognized that chromium salts can be leached from leather by lactic acid and the lactate portion of sweat (Roddy and Lollar[269]) and that leather was tanned by trivalent chromium in the one-bath method. From this information, he concluded that shoe-leather dermatitis could be caused by sensitization to the basic chromic sulfate leached from the shoe by the patient's sweat. He also recounted examples from the literature revealing that patients with suspected shoe-leather dermatitis had been tested with ammonium and potassium dichromate and chromium acetate, chloride, phosphate, carbonate, nitrate, oxide, and fluoride. The tests were not positive, but none of these compounds is found in shoe leather. His pronouncement launched an academic controversy that has brought forth much valuable information concerning the behavior of trivalent and hexavalent chromium in animal and human skin.

Two years later, Samitz and Gross[279] published their findings after using sweat to extract chromium from chromium-tanned leather. They verified the work of Roddy and Lollar and recovered chromium from all the leather samples tested. However, in their tests with leather, both trivalent chromium and hexavalent chromium were leached by sweat and demonstrated by polarographic techniques. The leather samples tested showed considerably more trivalent than hexavalent chromium, but the authors concluded that the more diffusible hexavalent chromium was also more readily leached from leather by sweat. They agreed that trivalent chromium was present in leather, but they held that the presence of hexavalent chromium precluded dismissal of its role as a sensitizer. They theorized that the hexavalent chromium found in the samples could have been a contaminant in the preparation of the trivalent chromium used for tanning or an oxidation product of trivalent chromium brought about by some readily reducible substance present in the leather or by some agent added during the tanning process. Later, they showed[284] that the trivalent chromium extracted from leather by sweat was present as basic chromic sulfate complex, and the hexavalent chromium, as chromate ion. Samitz and Gross,[278] in later studies, patch-tested 12 chromate-sensitive patients with basic chromic sulfate and chromium nitrate. None of the patients reacted to the trivalent sulfate or nitrate. The authors concluded that there was no antigenic similarity between hexavalent and trivalent chromium. Continuing with this thesis, they directed their attention to a substance that could reduce hexavalent chromium to its trivalent form without harming the skin.[280] The agent developed was a mixture of sodium pyrosulfite (for reducing the chromium) and tartaric acid (for chelating the reduced trivalent chromium). An antichromium ointment containing these materials was an effective reaction blocker in two chromate-sensitive patients when it was applied 15 and 30 min after contact with potassium dichromate. An antichromium solution was somewhat less efficient.

At about the time of the studies of Samitz *et al.*, Cohen[63] became interested in the experimental production of circulating antibodies to chromium. He joined human red blood cells to hexavalent and trivalent chromium to form conjugated antigens. These conjugates were injected into rabbits to stimulate production of antiserum containing specific antibodies to chromium. The serum of the rabbits immunized with red cells sensitized to potassium dichromate did not contain antibodies against potassium dichromate. However, the serum immunized with red cells sensitized to chromium chloride did contain antibodies. Cohen concluded from the experiment that the immunized rabbits contained specific antibodies to trivalent chromium. He explained the results as follows: Polysaccharides are known to be capable of reducing hexavalent to trivalent chromium; the polysaccharides in skin (except the stratum corneum) may act as reducing agents for chromates that penetrate the ground substance; thus, externally applied chromate may penetrate the stratum corneum without forming any complexes in this layer, whereas at a deeper level it undergoes reduction; the resulting trivalent chromium then combines with polysaccharides or with protein to form a complex that stimulates antibody formation; externally applied trivalent chromium combines avidly with stratum corneum protein, but a chromate-sensitive patient will not react, because he has no antibodies to this complex.

In 1962, Mali *et al.*[200] conducted experiments that provided considerable information on the behavior of chromium compounds in skin. They performed experiments dealing with diffusion, measurement of membrane potential, capacity of chromium salts to bind with serum and dermal proteins, reduction of dichromate by skin components, permeation of chromium salts through living skin, tests on chromium-sensitive patients, and animal sensitization. From these experiments, they made several informative conclusions: Trivalent chromium has a strong affinity for epithelial and dermal tissues; hexavalent chromium has a small affinity for these structures; the affinity of trivalent chromium for epithelial and dermal structures reduces the diffusibility of the substance and therefore the ability to sensitize readily; the trivalent compounds developed from the reduction of hexavalent chromium would be excellently disposed to form haptene–protein complexes and initiate sensitization; the slight affinity of hexavalent chromium for tissue and its ability to penetrate cell membranes are conducive to a reaction in chromium-sensitive people; and the fact that guinea pigs can be sensitized for dichromate sensitivity by the intracutaneous injection of trivalent chromium makes it possible that a chromium–protein complex outside the cells is the initial step in sensitization.

Another example of the unusual implications of the chromium sensitivity question was reviewed by Cairns and Calnan[51] in their paper dealing with green tattoo reactions associated with cement dermatitis. They observed two patients who developed dermatitis from cement and reactions to the green color in tattoos. Both patients were sensitive to hexavalent chromium. Cement contains hexavalent chromium, whereas tattoo green (if a chromium pigment is used) contains trivalent chromium. Because both patients reacted to hexavalent-chromium patch tests and both showed activation of the tattoo site, the authors proposed that the trivalent tattoo pigment reacted because it was oxidized in tissue to the hexavalent form. The work of several others who had previously investigated tattoo reactions, particularly the detailed study of green tattoos, was reviewed by Loewenthal,[190] who concluded that the reaction to tattoo green was probably the result of contamination of trivalent chromium with some hexavalent material.

In 1963, Fonseca[104] published a monograph on dermatoses caused by chromium. The material included observations on chemical action, biologic effects, experimental cutaneous sensitization, and a clinical study of 243 workmen engaged in seven different professional activities involving chromium exposure. Much of his laboratory and clinical information agreed with what had been observed by others in Europe and in the United States.

In 1963, Samitz and Katz[281] conducted an extensive study of the chemical reactions between chromium and some components of skin and demonstrated that hexavalent chromium is reduced by skin to the trivalent form. Methionine, cystine, and cysteine can bring about this reduction. Similarly, hemoglobin and gamma globulin reduced hexavalent chromium. Only trivalent chromium was bound to skin; thus, hexavalent chromium must be reduced to trivalent chromium before binding can occur. They drew no conclusions concerning the antigenic mechanism, other than to state their opinion that the trivalent material was not antigenic. About a year later, Fregert and Rorsman[108] used intracutaneous tests to show that a person allergic to hexavalent chromium is regularly allergic to trivalent chromium (as 0.5M chromium trichloride). They noted that allergy to trivalent chromium had been observed in Europe by Bockendahl[40] in 1954, Skog[316] in 1955, Spier *et al.*[320] in 1956, and Zina[372] in 1956, and in the United States by Morris[224] in 1958. Several others were cited as having observed its occurrence in studies with guinea pigs and (in one instance) rabbits. Fregert and Rorsman discovered that patch tests with the trivalent material were not always positive in patients sensitive to hexavalent chromium. Consequently, their 22 known chromium-sensitive subjects were tested with intracutaneous injections of 0.01M chromium trichloride. An inflammatory response resulted in all the subjects. When 0.001M chromium trichloride was used, 12 of the 22 reacted. Patch tests with 0.5M chromium trichloride caused positive reactions in 11 of 17, and patch tests with 0.07M chromium trichloride caused positive reactions in four of 22. They concluded that allergy to hexavalent chromium unequivocally implies allergy to trivalent chromium, on the basis of intracutaneous and patch tests. They emphasized the need to use a greater concentration of trivalent test material than is needed in testing with hexavalent chromium, because the trivalent forms are less soluble and less absorbable.

It now appears that investigators agree on several points:

1. People who work with hexavalent chromium can develop cutaneous and nasal mucous-membrane ulcers, whereas exposure to trivalent chromium does not produce these effects.
2. People who work with hexavalent chromium compounds can develop contact dermatitis from these agents, and they react to patch and intracutaneous tests with nonirritant concentrations of potassium dichromate.
3. Hexavalent chromium in tissue is reduced to the trivalent form.
4. Hexavalent chromium has greater diffusibility and solubility in tissue than trivalent chromium.
5. Hexavalent chromium can readily penetrate membranes.
6. Trivalent chromium can readily bind with some proteins to form complexes.

However, there is no unanimity that sensitivity to hexavalent chromium unequivocally implies sensitivity to trivalent chromium or that trivalent chromium compounds generally cause positive reactions to patch or intracutaneous tests in persons sensitive to hexavalent chromium.

SUMMARY

Cutaneous injury from chromium has been known since 1827. Chromium reactions of skin are generally classified as primary irritant effects (which include corrosive ulcers, scars, and nonulcerative contact dermatitis) and allergic effects (eczematous and noneczematous contact dermatitis).

The chromium compounds best known for their ulcerogenic action are strong concentrations of chromic acid, sodium and potassium chromate and dichromate, and ammonium dichromate. Less concentrated hexavalent compounds cause allergic contact dermatitis.

It is known that hexavalent chromium is reduced within the skin to the trivalent state by methionine, cystine, and cysteine. It has been suggested, but not proved, that the trivalent form resulting from the reduction may form haptene–protein complexes and thereby initiate sensitization. Some investigators believe that all people sensitized to hexavalent chromium are also sensitive to the trivalent form.

Several antichromium agents—for example, 5% sodium hyposulfite; a mixture of sodium pyrosulfite, tartaric acid, glucose, and ammonium chloride; EDTA; or 10% aqueous ascorbic acid—can be used to prevent chromium ulcers or facilitate their healing.

8

Experimental Exposures of Animals to Chromium Compounds

PULMONARY REACTIONS

In the early experiments in which animals were exposed to chromium compounds, lung cancer was not mentioned, although other pulmonary effects were noted. Lehmann[185] in 1914 exposed one cat and one rabbit for several hours per day to dichromates in the air at concentrations (as chromium) of 4–8.5 mg/m^3. The cat died of bronchial pneumonia, but the rabbit suffered no lung damage during a month's exposure. A report in 1935 stated that inhalation of "chromic acid vapor" caused inflammatory reactions in the lungs of guinea pigs.[21] In an experiment by Lukanin,[194] a few animals were placed in a chromate-chemical manufacturing plant for 1–8 months or were subjected to comparable exposure in the laboratory. Diffuse thickening of the alveolar walls and proliferation of the cells along the blood vessels and bronchi were observed, but no tumors were mentioned. In contrast with these irritant effects resulting from exposure to hexavalent chromium, the inhalation of the trivalent compound, chromic oxydicarbonate, $Cr_2O(CO_3)_2$, as a powder at a concentration of 58 mg/m^3 produced no pathologic effects in the lungs of cats exposed repeatedly over a period of 4 months.[3]

Because of the recognition of lung cancer as a hazard to employees in the old chromate-manufacturing plants during the 1930's, a number of investigations were undertaken in an attempt to produce cancer of the lungs and other tissues in animals directly with various chromium chemicals (see Tables 7-12, 8-1, and 8-2).

Shimkin and Leiter[308] injected chromite ore intravenously into tumor-susceptible mice. Although chronic irritation resulted, no increase was observed in the incidence of primary pulmonary tumors compared with that in control mice; nor did the injection of the ore affect the development of tumors induced by the intravenous injection of methylcholanthrene. Schinz and Uehlinger[290,291] found three (possibly four) sarcomas in various areas of the body after the implantation of pure metallic chromium in the femora of rabbits.

Hueper[150] injected powdered metallic chromium and chromite ore into the femoral bone marrow, peritoneal cavity, pleural space, paranasal sinuses, skeletal muscle, and vascular lumen of mice, rats, guinea pigs, rabbits, and dogs. Although tumors of various types were found at autopsy, Hueper concluded that none of these appeared to be causally related to the implanted material. In later investigations, Hueper and Payne[151,153,154,241,242] performed a number of experiments with rats, using chromium chemicals that varied in solubility from the relatively insoluble chromite ore and metallic chromium to the highly soluble sodium dichromate and chromium acetate. These materials were combined with sheep fat, tricaprylin, or gelatin to maintain the contact of the chemical at a specific site in the tissue for a longer period. The materials were implanted chiefly into the pleural cavity and thigh muscle. Pathologic studies were conducted on all animals. No tumors developed with sodium chromate or chromium acetate. A number of sarcomas developed at the site of implantation when the moderately

TABLE 8-1 Analysis of a Typical Pulverized Roast Material Used in Experiments[a]

Component	Fraction, %
CrO_3	13.7
Na_2O	9.3
Cr_2O_3	6.9
Fe_2O_3	17.7
Al_2O_3	9.4
MgO	8.7
CaO	31.0
V_2O_5	0.2
SiO_2	2.4
Other	0.7
TOTAL	100.0

[a]After Baetjer *et al.*[20] For the inhalation exposures, 1% potassium dichromate was added to the mixed chromium material.

TABLE 8-2 Composition of Potassium Zinc Chromate[a]

Formula:	$K_2O \cdot 4ZnO \cdot 4CrO_3 \cdot 3H_2O$
Solubility in water:	Slightly soluble, with partial decomposition to a less soluble product, releasing potassium chromate and potassium dichromate into solution
Analysis:	CrO_3–45.4%
	ZnO –37.3%
	K_2O –10.7%
	H_2O – 6.5%

[a]After Baetjer *et al.*[20]

soluble compounds were tested—chromite ore roast, roast residue, sintered chromium trioxide,* sintered calcium chromate,* and calcium, zinc, strontium, and lead chromates. In some cases, the sarcomas were invasive. The incidence of sarcomas varied, depending on the specific compound and the tissue. Because many different materials produce sarcomas, these experiments cannot be considered definitive for chromium. Of much greater significance was the production of three squamous-cell carcinomas in the lungs of rats that received intrapleural injections—two of 25 rats with chromite ore roast and one of 35 rats with calcium chromate. The chemicals were mixed with sheep fat. An adenocarcinoma of the lung also was produced in one of 39 rats by repeated intrapleural injection of sodium dichromate in gelatin. Similar experiments in mice with sintered chromium trioxide, calcium chromate, and sintered calcium chromate were essentially negative. In another experiment from the same laboratory, rats were exposed by inhalation for 2 years to chromium metal, chromite ore, chromite ore roast, chromic oxide, or zinc, barium, or lead chromate; the results were negative. Intratracheal injection of calcium, strontium, or zinc chromate yielded only three sarcomas in 218 rats. When the roast residue was given with benzo[a]pyrene, the incidence of spontaneous tumors was lower than when benzo[a]pyrene was administered alone. Hueper and Payne concluded that the two factors crucial for the production of tumors in their experiments were the moderate solubility of their chromium compounds and the prolonged local contact with the tissue obtained by the use of a vehicle like sheep fat. Chromium compounds of higher and lower solubility were much less effective.

Baetjer *et al.*[20] also conducted a large series of experiments with vari-

*Heated at about 1100 C for 1 hr to reproduce conditions of the roast process.

ous species of animals, chromium chemicals, and routes of exposure. One mixed strain of rats and three strains of inbred mice—A, Swiss, and C57BL, with high, moderate, and low incidences of spontaneous lung tumors, respectively—were exposed by inhalation to a mixed chromium material comparable with that found in the air of a chromate-producing plant (roast material plus 1% potassium chromate, $K_2Cr_2O_4$). The median particle diameter was 0.8 μm, and the concentration of the soluble fraction (90% water-soluble and 10% acid-soluble) varied from 1 to 3 mg/m^3 (as chromium trioxide). Only 3% of the airborne chromium dust was insoluble in water or acid. The animals were exposed for 4 hr/day, 5 days/week for periods of 20–54 weeks. In other experiments, the animals were subjected to repeated intratracheal or intrapleural injection of this material suspended in olive oil or to intratracheal and intravenous injections of the highly insoluble compounds, potassium zinc chromate and barium chromate. Control animals received injections of olive oil, zinc carbonate, barium sulfate, or saline solution. Histologic sections of all abnormal areas in the lung were made. No bronchogenic carcinomas appeared in any of the experimental animals. Pulmonary adenomas occurred at an earlier age in the experimental mice than in their controls when they were exposed to the mixed dust by inhalation or intraperitoneal injection or to potassium zinc chromate by intratracheal or intravenous injection. Epithelialization of the alveoli developed in mice after repeated intratracheal injection of potassium zinc chromate.

Steffee and Baetjer[322] repeated the inhalation and intratracheal experiments just described, using rabbits, guinea pigs, rats, and mice. No malignant tumors and no greater incidence of adenomas of the respiratory tract were produced. One rat exposed by inhalation developed a keratinizing tumor of the lung, but it was judged as benign. The intratracheal injection of potassium zinc chromate produced a higher incidence of alveolar hyperplasia in rabbits, guinea pigs, and mice. Alveolar and interstitial inflammation and granulomata resulted from some exposures. Attempts to induce malignancy by combining chromium exposure with PR 8 influenza virus infection or with methylcholanthrene were unsuccessful.

Laskin, Kuschner, and Drew[179,181] have developed a method that provides a continuous effective dose of a chemical over a long period to one small area of the respiratory tract of rats. A very small capsule made of fine steel wire mesh is impregnated with the chemical, mixed in equal parts with cholesterol. The capsule is attached permanently with a fishhook to the inner wall of the lower-lobe bronchus in such a manner as not to cause obstruction. Using this method, the authors have tested the carcinogenic properties of the roast residue, calcium chromate, chromic chromate, chromic oxide, and chromic trioxide.

Six of 100 rats exposed to calcium chromate developed squamous-cell carcinoma of the bronchus, and two additional rats developed adenocarcinoma–all at the site of impaction. One of 100 rats exposed to the roast residue developed squamous-cell carcinoma. All other compounds were without effect. The tumors were invasive, and some metastasized. The mean duration of exposure for the six carcinomas caused by calcium chromate was 540 days, and the durations for the two adenocarcinomas were 366 and 609 days. The authors suggested that the irritant properties of the chromates and the prolonged dosage resulting from the moderate solubility of the chemicals played a role in the success of these experiments.

Stimulated by the experiments of Laskin *et al.*,[182] Roe and Carter[270] attempted to produce cancer in rats by repeated intramuscular injections of calcium chromate in arachis oil over a 20-week period. Sarcomas developed at the site of injection that were invasive but did not metastasize.

The possibility that chromium may operate as a cofactor in the production of cancer was tested by Nettesheim *et al.*[231] C57BL/6 mice were pretreated with 100 R of whole-body x radiation or infected with PR 8 influenza virus. Later inhalation of chromium oxide dust for 6–18 months had no measurable effect on tumor incidence. The only tumors observed were adenomas or adenocarcinomas. In a later paper,[230] these authors reported experiments with mice exposed for 5 hr/day, 5 days/week for their lifetime to fine calcium chromate dust (70% of particles smaller than 0.3 μm) at a concentration of 13 mg/m^3. The mice developed pulmonary adenomas at an earlier age than control mice and had an incidence four times as great. The control mice were exposed to filtered air. Pre-exposure to x radiation did not affect the tumor rate, but preceding PR 8 influenza infection reduced the chromate tumor incidence. No bronchogenic tumors occurred.

In summary, experiments with rats demonstrated that invasive sarcomas can be produced locally by the implantation of a relatively insoluble hexavalent chromium compound in a medium that holds the chemical in contact with specific areas of the pleural cavity and thigh muscle. In tumor-susceptible mice, adenomas occurred earlier and in some experiments at a higher incidence after exposure to some hexavalent chromium chemicals. However, the relation of these data to the problem of squamous-cell carcinoma of the bronchus in man is not clear. Of far greater significance has been the recent production of squamous-cell carcinoma of the bronchus in rats by the local implantation of calcium chromate in cholesterol pellets. Similarly, a few squamous-cell carcinomas have been produced in the lungs of rats by intrapleural injection of calcium chromate or a mixed material containing relatively insoluble hexavalent chromates in sheep fat. No dose–response relation is available from the current data. At present, although calcium chromate has been the specific chromium compound responsible for the bronchogenic carcinomas produced in rats, it would be premature to conclude that only this form of chromium can exert a carcinogenic effect.

The available data suggest that trivalent chromium chemicals do not play a role in the production of bronchogenic carcinoma.

SYSTEMIC REACTIONS

In addition to the pulmonary studies just described, systemic reactions to chromium administration have been examined in animals. Addition of moderate amounts of trivalent chromium to food or water did not have any harmful effect. For example, cats fed chromic phosphate or oxydicarbonate at 50–1,000 mg/day for 80 days were unaffected.[3] Similarly, no toxic reactions were observed in rats when their drinking water contained 25 ppm for a year[198] or 5 ppm throughout their lifetime.[296,298] The toxicity of trivalent chromium is so low that, even by parenteral administration, chromic acetate at 2.29 g/kg or chromic chloride at 0.8 g/kg is required to kill mice.[297] Even very large doses given intragastrically were not fatal to dogs. Brard[43] reported that 10 or 15 g of chromium as chromic chloride caused vomiting and diarrhea, but that recovery was rapid. However, the subcutaneous administration of 8 g of chromium as chromic chloride proved fatal in one dog. Some fatal doses of trivalent chromium compounds reported in the literature are listed in Table 8-3.

TABLE 8-3 Fatal Doses of Trivalent Chromium in Animals

Animal	Route[a]	Material	Chromium Dose, g/kg	Effect	Reference
Dog	SC	Chromic chloride	0.8	Fatal	43
Rabbit	SC	Chromic chloride	0.52	Fatal	43
Rat	IV	Chrome alum Cr-hexaurea chloride	0.01–0.018	LD_{50}	215
Mouse	IV	Chromic chloride	0.8	MLD	331
Mouse	IV	Chromic acetate	2.29	MLD	331
Mouse	IV	Chromic chloride	0.4	MLD	292
Mouse	IV	Cr^{+3} (?)	0.25–2.3	MLD	297
Mouse	IV	Chromic sulfate	0.085	MLD	292
Mouse	IV	Chromium carbonyl	0.03	LD_5	292

[a]SC, subcutaneous; IV, intravenous.

TABLE 8-4 Effects of Hexavalent Chromium in Animals[a]

Animal	Route[b]	Material	Average Dose or Concentration	Duration	Effect	Reference
Rabbit and cat	IH	Chromates	1–50 mg/m^3	14 hr/day for 1–8 mo	Pathologic changes in lungs	194
Rabbit	IH	Dichromates	11–23 mg/m^3 as dichromate	2–3 hr/day for 5 days	None	185
Cat	IH	Dichromates	11–23 mg/m^3 as dichromate	2–3 hr/day for 5 days	Bronchitis, pneumonia	185
Mouse	IH	Mixed dust containing chromates	1.5 mg/m^3 as CrO_3	4 hr/day, 5 days/wk for 1 yr	None harmful	20,322
Mouse	IH	Mixed dust containing chromates	16–27 mg/m^3 as CrO_3	½ hr/day intermittently	Fatal to some strains	20,322
Mouse	IH	Mixed dust containing chromates	7 mg/m^3 as CrO_3	37 hr over 10 days	Fatal	20,322
Rat	IH	Mixed dust containing chromates	7 mg/m^3 as CrO_3	37 hr over 10 days	Barely tolerated	20,322
Rabbit and guinea pig	IH	Mixed dust containing chromates	5 mg/m^3 as CrO_3	4 hr/day, 5 days/wk for 1 yr	None marked	20,322
Rat	OR	Potassium chromate in drinking water	500 ppm	Daily	Maximal nontoxic concentration	125
Mature rat and mouse	OR	Zinc chromate in feed	1%	Daily	Maximal nontoxic concentration	125
Young rat	OR	Zinc chromate in feed	0.12%	Daily	Maximal nontoxic concentration	125
Young rat	OR	Potassium chromate in feed	0.12%	Daily	Maximal nontoxic concentration	125
Dog, cat, and rabbit	OR	Monochromates or dichromates	1.9–5.5 mg chromium/kg body wt per day (1 mg chromium equivalent to 2.83 mg $K_2Cr_2O_7$ or 3.8 mg K_2CrO_4)	29–685 days	None harmful	185

[illegible]	[illegible]	Potassium dichromate	1–2 g as Cr	Daily	Fatal in 3 months	43
Dog	ST	Potassium dichromate	1–10 g as Cr	–	Rapidly fatal	43
Monkey	SC	Potassium dichromate	0.02–0.7 g in 2% solution	–	Fatal[c]	155
Dog	SC	Potassium dichromate	210 mg as Cr	–	Rapidly fatal	43
Guinea pig	SC	Potassium dichromate	10 mg	–	Lethal[c]	237,238
Rabbit	SC	Potassium dichromate	1.5 cc of 1% solution/kg body wt	–	80% fatal[c]	137
Rabbit	SC	Potassium dichromate	20 mg	–	Lethal[c]	233
Rabbit	SC	Potassium dichromate	0.5–1 cc of 0.5% solution/kg body wt	–	Nephritis[c]	233
Rabbit and guinea pig	SC or IV	Sodium chromate	0.1–0.3 g as CrO_3	–	Rapid death[c]	256
Mouse	IV	Zinc chromate	0.1 mg/month as zinc chromate	10 mo	Tolerated	256
Mouse	IV	Zinc chromate	0.75 mg	1 dose	Fatal	256
Mouse	IV	Barium chromate	2.5 mg/dose as barium chromate	9 doses at 6-wk intervals	Tolerated	256
Rabbit	IV	Potassium dichromate	0.7 cc of 2% solution/kg body wt	–	Fatal[c]	207
Dog	IV	Potassium chromate	10 grains	–	Fatal	114
Dog	IV	Potassium chromate	1 grain	–	Survived	114
Dog	IV	Potassium dichromate	210 mg as Cr	–	Rapidly fatal	43
Dog	IV	Potassium dichromate	3 mg/100 cc blood per dose	2 doses	Marked renal damage	139,314

[a]After Baetjer.[18]
[b]IH, inhalation; OR, ingestion; ST, stomach tube; SC, subcutaneous; IV, intravenous.
[c]Renal damage.

Hexavalent chromium chemicals can be tolerated by animals in low concentrations, especially when they are administered in feed or drinking water, in which the degree of absorption is a factor. For example, rats tolerated hexavalent chromium in drinking water at 25 ppm for a year,[198] and dogs showed no effect of chromium as potassium chromate at 0.45–11.2 ppm over a 4-year period.[11] Even higher concentrations have been reported by some investigators. However, larger doses of hexavalent chromium are highly toxic and may cause death, especially when injected intravenously, subcutaneously, or intragastrically.[18] The fatal doses of hexavalent chromium compounds by route of administration, as reported in the literature, are shown in Table 8-4. Intragastric administration results in marked irritation and corrosion of the gastrointestinal tract. In dogs, vomiting, diarrhea, gastric and intestinal hemorrhage, and death followed rapidly.[43] Kidney lesions are the most common type of systemic damage in all species, especially if the chromates are injected subcutaneously or intravenously. Nephritis that involves chiefly the tubules, with destruction of the epithelium, has been described by many of the early experimenters in this field.[137,139,155,207,233,237,314] Kidney damage in animals also has been reported after absorption of hexavalent chromium from mucous membranes of the digestive tract. In dogs, the application of a solution of potassium dichromate to the oral mucosa (30 mg/kg) caused not only a local corrosive action, but also impairment of kidney function and death. Intragastric administration at 5.2 mg/kg produced the same result. However, the application of a 2% chromium trioxide solution to the skin had only local effects.[313]

Thus, the systemic toxic effect of chromium on animals depends on the valence of the chromium. Trivalent chromium is poorly absorbed and has a low degree of toxicity. Hexavalent chromium is irritating and corrosive to the mucous membranes, is absorbed more readily, and is highly toxic when introduced systemically.

9

Effects of Chromium on Vegetation

Very little information is available on the effects of chromium on plants. In general, low concentrations of chromium in water or soil appear to be beneficial or possibly even essential to plants, whereas higher concentrations may be toxic. The effects vary with the species and with the specific chromium compound.

In water plants, hexavalent chromium at 0.03–64 ppm inhibits the growth of algae, whereas lower concentrations stimulate growth in some cases. Hexavalent chromium at 1–5 ppm in seawater reduces photosynthesis of giant kelp.[306]

In the case of land plants, the effects on growth of adding chromium to the soil depend on the amount of chromium naturally present in the soil. Crop yields have been improved by application of chromium to soils in Germany, France, Poland, and Russia.[210] The addition of chromic sulfate to soil at 600 g/ha (grams/hectare) improved the weight, size, and sugar content of grapes by 21, 18, and 23%, respectively, and increased the yield by 205–245 kg/ha.[82] The application of a fertilizer containing 0.43% chromium resulted in increased growth of flax grown on sand.[178] Chromic sulfate applied to mulberry leaves in a 1 : 10,000 solution increased cocoon and silk weight of silkworms grown on these leaves by 14–16.5%.[337] Addition of chromous acetate (0.05% or less) to soil had a beneficial effect on carrots, barley, lupines, and cucumbers.[169] Application of chromium (as the alum) at 40 g/ha to a soil containing extractable chromium at only 65 μg/kg increased the yield of potatoes from 32.7 to 46.5 tons/ha.[29] Similar results were obtained with peas, carrots, and beets.[27,28]

Pratt[255] recently reported that applications of potassium dichromate at 30 and 100 g/m^3 of soil increased the yield of cucumbers; chromium at 0.1 ppm in nutrient solutions benefited lettuce slightly; and chromium added at 5 mg/kg to soil increased the rate of nitrification.

However, Pratt[255] cited a number of observations of toxic effects of chromium at higher concentrations. Although chromium at 75 ppm in soil was not harmful to orange seedlings, the addition of chromium at 150 ppm was toxic. Chromic sulfate stimulated the growth of corn seedlings in culture solutions containing chromium at 0.5 ppm, but at 5 ppm and above it inhibited growth. The growth of tomatoes, oats, kale, and potatoes was reduced by chromium (as chromate) at 16 ppm. Chromium at 5 and 10 ppm in nutrient solutions produced iron chlorosis in oat plants, and at 15–50 ppm it was toxic. Chromium (chromic or chromate) at 8 and 16 ppm produced iron chlorosis in sugar beets, and at 5 ppm (as chromate) it was toxic to tobacco and at 10 ppm toxic to corn.

In some instances, toxicity has been associated with the chromium concentration in the plant tissues. For example, tobacco leaves grown on serpentine soil, which normally has a high chromium concentration (possibly several percent), may contain chromium at 14 ppm (dry wt) without toxic signs; but at 18–34 ppm, toxic effects were visible. Concentrations of 175 ppm (dry wt) in the roots were without harm; but at 375–410 ppm, toxic symptoms were present. In fruits, vegetables, and grain, no harmful evidence was found with concentrations from traces to about 14 ppm (dry tissue); but toxic symptoms appeared in corn when the leaves contained 4–8 ppm and in oats when the leaves contained 252 ppm.[255]

The effect on the surrounding vegetation of chromium discharged into the air from industrial sources has been observed only rarely. In the neighborhood of a Swiss chrome-plating factory, which discharged hexavalent chromium, burning and necrosis of vegetables and fruit and ornamental trees were noted in a commercial garden. Except for the leafy vegetables, those plants regained their vigor later in the growing season. The concentration of chromium in the plant tissue varied from less than 1 to 9.8 μg/g of tissue. The chromium content of the soil was 8.4–30 μg/g at the surface and 30–71 μg/g at a depth of 30 cm. In contrast, control areas had concentrations ranging 1.1–1.9 μg/g.[81] In the region of a Swedish ferroalloy plant, the chromium concentration in the moss *Hypnum cupressiforme* ranged as high as 12,000 ppm, compared with a normal value of 10 ppm.[276] No damage to the moss was reported. According to M. Piscator (personal communication), the uptake of metals by moss is independent of their concentrations in soil and soil water. Moss grows mainly on rocks, and its intakes are entirely from deposited material. Moss acts as an ion exchanger and can accumulate large amounts of metal without injury.

One acute episode of plant damage resulting from the accidental discharge of sodium dichromate was reported in Japan in 1969. About 2 kg of the material was spread over an area within a radius of 0.2 km from the plant, producing an air concentration of 50–150 $\mu g/m^3$. The leaves of rice plants showed brown to black spots attributed to concentrations of about 0.85–72 ppm in the tissue.[205]

In summary, it appears that chromium exerts its effects, both beneficial and toxic, on the roots of plants; thus, the concentration of available chromium in the water and soil is the determining factor. In evaluating the role of chromium in water and soil, the chemical form of the chromium and its solubility and concentration are the important factors. In addition, consideration must be given to the presence of other chemicals in the soil that may interact with chromium in its effects on vegetation—notably nickel, cobalt, and magnesium. There are no data from which the dose–response relation of airborne chromium to vegetation can be determined; in fact, no data were found from which to determine whether airborne chromium exerted a direct effect on plants apart from an indirect contribution of air contamination to the soil and water. The contribution of airborne chromium to water and soil seems to be inappreciable under normal circumstances. Natural soil usually contains chromium at 5–3,000 ppm (mean, 40 ppm), and soil derived from ultrabasic or serpentine rocks contains even greater concentrations (see Table 2-1, p. 9). The chromium content of soil may be increased by treatment with superphosphate that contains chromium at 66–243 ppm and by the application of organic fungicides and other chemicals.[22] Thus, the concentrations of chromium usually present in air (traces to 0.02 $\mu g/m^3$ as Cr) are too low to have any significant effect on the growth or yield of vegetation.

10

Chromium in Aquatic Species

The Panel does not have the necessary expertise to discuss the effects of chromium on aquatic species. However, the NAS–NAE Committee on Water Quality Criteria is publishing a report[228] that discusses the effects of chromium on marine and freshwater aquatic species. The Panel presents these data without evaluation or critical analysis. The concentrations of chromium in a few aquatic species of plants and animals, as cited in other sources, are listed in Tables 10-1 through 10-3. Sufficient data are not available to evaluate the role of chromium in the ecologic system.

The concentrations of chromium in both seawater and fresh water that may have effects on aquatic species have been reviewed by the Committee on Water Quality Criteria. The portion of its report on seawater includes the following statement:[228]

> Chromium concentrations in seawater average about 0.04 μg/1 (Food and Agriculture Organization 1971),[339*] and concentration factors of 1,600 in benthic algae, 2,300 in phytoplankton, 1,900 in zooplankton, 440 in soft parts of molluscs, 100 in crustacean muscle, and 70 in fish muscle have been reported (Lowman et al. 1971).[193]
>
> The toxicity of chromium to aquatic life will vary with valence state, form, pH,

*Reference numbers in this extract have been changed to conform with this volume.

synergistic or antagonistic effects from other constituents, and the species of organism involved.

In long-term studies on the effects of heavy metals on oysters, Haydu (*unpublished data*) [Weyerhaeuser Company] showed that mortalities occur at concentrations of 10 to 12 μg/1 chromium, with highest mortality during May, June, and July. Raymont and Shields (1964)[263] reported threshold toxicity levels of 5 mg/1 chromium for small prawns (*Leander squilla*), 20 mg/1 chromium in the form Na_2CrO_4 for the shore crab (*Carcinas maenus*), and 1 mg/1 for the polychaete *Nereis virens*. Pringle et al. (1968)[258] showed that chromium concentrations of 0.1 and 0.2 mg/1, in the form of $K_2Cr_2O_7$, produced the same mortality with molluscs as the controls. Doudoroff and Katz (1953)[84] investigated the effect of $K_2Cr_2O_7$ on mummichogs (*Fundulus heteroclitus*) and found that they tolerated a concentration of 200 mg/1 in sea water for over a week.

Holland et al. (1960)[144] reported that 31.8 mg/1 of chromium as potassium chromate in sea water gave 100 per cent mortality to coho salmon (*Oncorhynchus kisutch*). Gooding [1956][117] found that 17.8 mg/1 of hexavalent chromium was toxic to the same species in sea water.

Clendenning and North (1960)[62] showed that hexavalent chromium at 5.0 mg/1 chromium reduced photosynthesis in the giant kelp (Macrocystis) by 50 per cent during 4 days exposure.

The Committee on Water Quality Criteria also discussed freshwater aquatic species.[228] The following statements are taken from that document:

TABLE 10-1 Chromium in Aquatic Plants and Animals[a]

Type of Organism	Chromium Concentration, ppm (dry wt)
Plants	
Plankton	3.5
Brown algae	1.3
Bryophytes	2
Ferns	0.8
Gymnosperms	0.16
Angiosperms	0.23
Fungi	1.5
Animals	
Coelenterates	1.3
Mollusks	–
Echinoderms	0.5
Crustaceans	–
Insects	–
Fish	0.2
Mammals	0.3

[a]Derived from Schroeder.[292]

TABLE 10-2 Chromium in Shellfish from Atlantic and Gulf Coast Regions, 1965–1970[a]

Type of Shellfish	No. Samples	Chromium Concentration, ppm (drained wet tissue)	
		Mean	Range
Hard clam	129	0.28	0.06–0.35
Soft-shell clam	59	1.62	1.09–1.83
Oyster	356	0.37	0.30–0.51
Surf clam	23	7.26	–

[a]Data from Division of Shellfish Sanitation, Office of Food Sanitation, Bureau of Foods, Food and Drug Administration (personal communication).

The chromic toxicity of hexavalent chromium to fish has been studied by Olson (1958)[234]* and Olson and Foster (1956, 1957).[235,236] Their data demonstrated a pronounced cumulative toxicity of chromium to rainbow trout and chinook salmon (*Oncorhynchus tshawytscha*). Doudoroff and Katz (1953)[84] found that bluegills (*Lepomis macrochirus*) tolerated a 45 mg/1 level for 20 days in hard water. Cairns (1956),[50] using chromic oxide (CrO_3), found that a concentration of 104 mg/1 was toxic to bluegills in 6 to 84 hours. Bioassays conducted with four species of fish gave 96-hour LC50's of hexavalent chromium that ranged from 17 to 118 mg/1 indicating little effect of hardness on toxicity (Pickering and Henderson 1966).[247]

Recently some tests of chronic effects on reproduction of fish have been carried out. The 96-hour LC50 and safe concentrations for hexavalent chromium were 33 and 1.0 mg/1 for fathead minnows (*Pimephales promelas*) in hard water (Pickering *unpublished data* 1971), 50 and 0.6 mg/1 for brook trout (*Salvelinus fontinalis*) in soft water, and 69 and 0.3 mg/1 for rainbow trout (*Salmo gairdneri*) in soft water (Benoit *unpublished data* 1971). Equivalent values for trivalent chromium were little different: 27 mg/1 for the 96-hour LC50, and 1.0 mg/1 for a safe concentration for fathead minnows in hard water (Pickering *unpublished data* 1971).

For *Daphnia* the LC50 of hexavalent chromium was reported as 0.05 mg/1, and the chronic no-effect level of trivalent chromium on reproduction was 0.33 mg/1 (Biesinger and Christensen *unpublished data* 1971). Some data are available concerning the toxicity of chromium to algae. The concentrations of chromium that inhibited growth for the test organisms are as follows (Hervey 1949):[143] Chlorococcales, 3.2 to 6.4 mg/1; Euglenoids, 0.32 to 1.6 mg/1; and diatoms, 0.032 to 0.32 mg/1. Patrick (*unpublished data* 1971) found that 50 per cent growth reduction for two diatoms in hard and soft water occurred at 0.2 to 0.4 mg/1 chromium.

Thus, it is apparent that there is a great range of sensitivity to chromium among different species of organisms and in different waters. Those lethal levels reported above are 17 to 118 mg/1 for fish, 0.05 mg/1 for invertebrates, and 0.032 to 6.4 mg/1 for algae, the highest value being 3,700 times the lowest one. The apparent 'safe' concentration for fish is moderately high, but the recommended maximum

*Reference numbers in this extract have been changed to conform with this volume.

concentration of 0.05 mg/1 has been selected in order to protect other organisms, in particular *Daphnia* and certain diatoms which are affected at slightly below this concentration.

TABLE 10-3 Chromium in Shellfish Harvested from Atlantic and Pacific Waters, United States[a]

	Chromium Concentration, ppm (wet wt)	
Type of Shellfish	Mean	Range
East Coast oyster	0.4	0.04–3.40
West Coast oyster	–	0.10–0.30
Soft-shell clam	0.52	0.10–5.0
Northern quahog	0.31	0.19–5.80

[a]Derived from Pringle *et al.*[258]

11

Standards for Chromium Content of Air and Water

OCCUPATIONAL EXPOSURE

United States

Limits suggested for the chromium concentration in the air of industrial plants have been in existence for a number of years and are currently published by three agencies—the American Conference of Governmental Industrial Hygienists (ACGIH),[7] the American National Standards Institute (ANSI),[9] and the United States Department of Labor under the 1970 Occupational Safety and Health Act (OSHA).[347] These limits are presented in Table 11-1.

THRESHOLD LIMIT VALUES

The ACGIH limits, called "threshold limit values" (TLV), were first published in 1947 and have been revised and enlarged yearly since then. The general concept embodied in the TLV's, as stated in the 1971 publication,[7(pp.1-3)] is reproduced in part below.

Threshold limit values refer to airborne concentrations of substances and represent conditions under which it is believed that nearly all workers may be repeatedly exposed day after day without adverse effect. Because of wide variation in individual susceptibility, however, a small percentage of workers may experience discomfort from some substances at concentrations at or below the threshold limit; a smaller percentage may be affected more seriously by aggravation of a pre-existing condition or by development of an occupational illness.

Threshold limit values refer to time-weighted concentrations for a 7 or 8-hour workday and 40-hour workweek. They should be used as guides in the control of health hazards and should not be used as fine lines between safe and dangerous concentrations. . . .

Threshold limits are based on the best available information from industrial experience, from experimental human and animal studies, and, when possible, from a combination of the three. The basis on which the values are established may differ from substance to substance; protection against impairment of health may be a guiding factor for some, whereas reasonable freedom from irritation, narcosis, nuisance or other forms of stress may form the basis for others.

The committee holds to the opinion that limits based on physical irritation should be considered no less binding than those based on physical impairment. There is increasing evidence that physical irritation may initiate, promote or accelerate physical impairment through interaction with other chemical or biologic agents.

TABLE 11-1 Standards for Chromium Exposures in Industry

	Standard, mg/m^3		
Source	ACGIH 1972[a]	ASA 1943, Reconfirmed ANSI 1971[b]	U.S. Dept. Labor OSHA 1972[c]
Chromic acid and chromates, as CrO_3	0.1	0.1	0.1
Chromium, soluble chromic and chromous salts, as Cr	0.5	–	0.5
Chromium metal and insoluble salts, as Cr	1.0	–	1.0
Some insoluble chromates	0.1 (1972 intended change)		

[a]ACGIH (American Conference of Governmental Industrial Hygienists):[7] "threshold limit values [TLV] refer to time-weighted concentrations for a 7 or 8-hour workday and 40-hour workweek." "When no ceiling value is given, time-weighted averages permit excursions above the limit provided they are compensated by equivalent excursions below the limit during the workday." For chemicals having a TLV of less than 1 mg/m^3, the maximal concentration permitted for short-time exposure is 3 times the TLV.

[b]ANSI (American National Standards Institute), formerly called ASA (American Standards Association):[9] "the maximum allowable concentration of chromium as chromate or dichromate dust, or as chromic acid mist, . . . for exposures not exceeding a total of eight hours daily."

[c]U.S. Dept. Labor, OSHA (Occupational Safety and Health Act):[347] for chromium, soluble chromic and chromous salts, metal, and insoluble salts, "an employee's exposure, . . . in any 8-hour work shift of a 40-hour work week, shall not exceed the 8-hour time weighted average limit given for that material in the table"; for chromic acids and chromates, the standard is the "acceptable ceiling concentration"; time-weighted averages and peaks over this value are not allowed.

In spite of the fact that serious injury is not believed likely as a result of exposure to the threshold limit concentrations, the best practice is to maintain concentrations of all atmospheric contaminants as low as is practical.

These limits are intended for use in the practice of industrial hygiene and should be interpreted and applied only by a person trained in this discipline. They are not intended for use, or for modification for use, (1) as a relative index of hazard or toxicity, (2) in the evaluation or control of community air pollution or air pollution nuisances, (3) in estimating the toxic potential of continuous uninterrupted exposures, (4) as proof or disproof of an existing disease or physical condition, or (5) for adoption by countries whose working conditions differ from those in the United States of America and where substances and processes differ.

The bases on which the TLV's have been determined have been published in the ACGIH's *Documentation of the Threshold Limit Values for Substances in Workroom Air.*[6] To evaluate the TLV's recommended for chromium and its compounds, the data on which they were based must be reviewed in detail.

1. *Documentation of TLV for Hexavalent Chromium as Related to Nasal Septal Perforation* The TLV for hexavalent chromium of 0.1 mg/m^3 (as CrO_3) originally proposed by the ACGIH for chromic acid mist was based on an examination of 23 workers in six chromium-plating plants conducted in 1928 by Bloomfield and Blum[39] (see Chapter 7). Studies of the ventilation methods used at the plating tanks

TABLE 11-2 Perforation of Nasal Septum in Chromate Workers[a]

Ratio of insol Cr^{+3} to sol Cr^{+6}	Chromium Concentration, mg/m^3 (as Cr)	No. Workers Examined	Workers with Septal Perforation	
			No.	%
Workers in plant				
≤ 1.0:1	≤ 0.25	4	2	50
	0.26–0.51	7	3	43
	≥ 0.52	8	4	50
1.1–4.9:1	≤ 0.25	9	7	78
	0.26–0.51	32	20	63
	≥ 0.52	15	11	73
≥ 5.0:1	≤ 0.25	7	2	29
	0.26–0.51	2	1	50
	≥ 0.52	13	11	85
TOTAL	–	97	61	63
Office workers	0.06	4	0	0

[a]Derived from Mancuso.[202]

demonstrated that it was feasible to reduce the air chromium content to less than 0.1 mg/m^3 (as CrO_3). Apparently on this basis, the authors concluded that "daily exposure to concentrations of chromic acid greater than 1 milligram in 10 cubic meters [0.1 mg/m^3] is likely to cause definite injury to the nasal tissues of the operators."[39] No observations were reported with concentrations at or below 0.1 mg/m^3. Although Bloomfield and Blum did not specify that 0.1 mg/m^3 constituted a safe exposure, it was adopted by the ACGIH for the TLV list, undoubtedly because no other data were available at that time.

In the 1971 *Documentation*, a number of other studies are cited as supporting evidence for industrial exposure to this concentration of chromic acid and chromates.

Mancuso,[202] in a study of an old chromate-producing plant, found 63% of the 97 workers with nasal septal perforation. To differentiate the roles of trivalent and hexavalent chromium, he divided the workers into groups according to the concentration and ratio of trivalent to hexavalent chromium in their exposure. The data are shown in Table 11-2. Nasal septal perforation occurred in all groups and was not consistently related to either the concentration or the trivalent:hexavalent ratio. Nasal septal perforation was found in several workers in a cement plant 1,000 ft downwind from the chromate plant where the ambient air concentration was 0.08 mg/m^3 (as Cr). Because no air analyses were made in this plant and cement contains chromium, the total exposure of these workers is not known.

In an extensive study conducted by the Public Health Service of all chromate-manufacturing plants in the United States in 1950,[97] perforation of the nasal septum was found in 56.7% of the 897 workers. The range of the means for the time-weighted average exposures in these plants was as follows:

- To soluble hexavalent chromium, 0.005–0.17 mg/m^3 (as Cr).
- To acid-soluble and water-insoluble chromium (includes small fraction of hexavalent chromium), 0–0.47 mg/m^3 (as Cr).
- To chromite ore, 0–0.89 mg/m^3 (as Cr).

Unfortunately, the concentrations associated with perforation of the septum were not specified.

In a more recent study of nine chrome platers exposed to chromic acid (CrO_3) mist at concentrations of 0.18–1.4 mg/m^3, the following conditions of the nasal septa were observed: Four men with 2–12 months of exposure had perforations; three men with 1–10 months of exposure had ulcers; and two men with ½–9 months of exposure had irritation.[167]

Finally, the documentation contains the following statement:[6 (p.55)] "Vigliani and Zurlo[354] reported ulcers of the nasal septum, irritation of the mucous membranes of the larynx, pharynx and conjunctiva, and chronic asthmatic bronchitis in a group of workers allegedly exposed to chromates or chromic acid in concentrations ranging from 0.11 to 0.15 mg/m^3." (See the section below, on Italian standards.)

2. *Documentation of TLV for Hexavalent Chromium as Related to Lung Cancer* The TLV for chromic acid and chromates apparently applies not only to perforation of the nasal septum but also to lung cancer, inasmuch as a number of references to lung cancer in chromate workers are cited in the documentation of TLV's.

Machle and Gregorius[197] in 1948 reported their study of lung cancer in six chromate-producing plants in the United States. The death rate due to cancer of the bronchi and lungs varied from 0 to 4.86 cases/1,000 males in the different plants. The duration of exposure in these cases varied from 4 to 47 years. The concentration of chromates in the air of four plants is shown in Table 11-3. The authors stated that the analytic data could not be used to evaluate the interplant differences in lung-cancer rates. Even within any one plant, tumor rates could not be associated with the concentration measured at a given task because of shifting of personnel and insufficient records of work assignments over the years of exposure.

Mancuso and Hueper[204] in 1951 measured the chromium exposure in a chromate-producing plant where seven cases of lung cancer had occurred in a plant population of 210. The data for the lung-cancer cases are shown in Table 11-4. The authors stated that the number of cases was too small to provide statistical correlation with the exposure years.

TABLE 11-3 Concentration of Chromates in Air by Location in Chromate-Producing Plants in United States[a]

Type of Location	Chromate Concentration, mg/m^3			
	Plant 1	Plant 2	Plant 3	Plant 4
Kilns and mills	0.06–1.00	0.30–2.80	0.80–4.60	0.01– 1.40
Dryers	0.20	0.04	0.34	0.25
Packing	0.20[b]	–	–	5.0 –21.0
General air	0.02–1.00	–	0.04	0.003
Concentrators	0.02–0.20	0.02	2.17	1.3 – 0.28
Granulators	0.03–0.57	0.01–0.73	–	0.12– 0.14

[a]Derived from Machle and Gregorius.[197]
[b]From undetectable to 0.20.

TABLE 11-4 Lung Cancer in Chromate Workers[a]

Case	Exposure Period, yr	Latent Period, yr	Weighted Average Exposure, mg/m^3 (as Cr)		
			insol	sol	Total
1	9.0	10.0	0.19	0.09	0.28
2	14.5	14.3	0.19	0.04	0.23
3	12.5	12.5	0.10	0.01	0.11
4	7.5	9.0	0.48	0.15	0.63
5	9.25	14.0	0.58	0.08	0.66
6	2.0	7.2	0.10	0.01	0.11
7	7.25	7.25	0.58	0.08	0.66

[a]Derived from Mancuso and Hueper.[204]

In 1953, the Public Health Service[97] reported a high incidence of lung cancer in the chromate-manufacturing plants in the United States. The mean weighted exposures found in this study are given above, but, as in the case of septal perforation, the specific exposures related to the cancer cases were not determined.

Bidstrup[34] and Buckell and Harvey[46] reported the occurrence of one case of bronchogenic carcinoma in 724 chromate workers in a British plant where the mean concentration of trivalent chromium varied from 0.0006 to 2.14 mg/m^3, with a maximum of 3.27 mg/m^3, and that of hexavalent chromium varied from 0.005 to 0.88 mg/m^3 (as Cr), with a maximum of 5.68 mg/m^3.

The only other publication[16] on human exposure to chromates cited in the TLV documentation reported the concentrations that existed in the old German chromate-producing plants where lung cancer was first noted. In the three plants cited, the concentrations were 0.1–0.5, 0.1–37.2, and 5–40 mg/m^3 (as Cr) for monochromates, and up to 50 mg/m^3 (as Cr) for dichromates. The specific exposure in cancer cases was not given in the German papers.

Stokinger[325] described the procedures used in validating some of the TLV's in a paper published in 1969. Regarding the chromates, he stated:

> To give a clearer idea of how these validation procedures work, a most productive day's meeting was held with industrial physicians and hygienists of the chromate industry. The reason for selecting the chromate industry was that no evidence of the suitability of the TLV for chromic acid or chromates had ever been brought forth for the prevention of either nasal perforation or bronchogenic carcinoma. All environmental levels had exceeded the recommended limit of 0.1 mg/cum when large excesses (29-fold) of lung cancer and nasal perforation were found in 1950, and the health experience had not been reviewed in these terms since the industry had improved its control measures to the recommended limit. In brief,

the day's discussion revealed that the limit for chromic acid mist was satisfactory in preventing nasal perforation, and in addition contained a safety factor of three or four; that the limit was probably satisfactory for the prevention of lung cancer, as no new cases have appeared since the reduction in exposure occurred, but that the ten years in which the closer controls were operative are probably too short a time to be certain of its validity in this respect.

The conclusions from this meeting were reproduced in the 1971 *Documentation.* No data to substantiate these conclusions were given in either the TLV documentation or Stokinger's paper.

The documentation for chromic acid and chromates concludes that "the evidence available indicates that concentration of CrO_3 mists and soluble chromate dusts must be kept close to 0.1 mg/m^3 (as CrO_3) to prevent irritation and injury to the nasal passages and other respiratory symptoms. To produce lung cancer it seems probable that higher concentrations, possibly of sparingly soluble chromates, are necessary."[6 (p.56)]

3. *Documentation of TLV for Trivalent Chromium* The TLV for soluble chromic and chromous salts is 0.5 mg/m^3 (as Cr), and that for chromium metal and insoluble salts, 1.0 mg/m^3 (as Cr). The documentation cites three types of data as the basis for these figures: experiments in which animals were exposed to trivalent chromium in air and water, cancer studies in rats, and observations on humans.

Akatsuka and Fairhall[3] exposed two cats for ½–1 hr/day for 4 months to chromic carbonate in concentrations of 67–140 mg/m^3 (as Cr). Ten cats were given chromium carbonate or phosphate in the water at 50–1,000 mg/day. No harmful effects were found either by pathologic examination or by chemical analysis of tissues. In another study,[198] rats given water containing hexavalent or trivalent chromium at 25 ppm for a year showed no toxic symptoms.

For the cancer problem, the work of Hueper and Payne[154] was cited. These investigators implanted 25 mg of chromic acetate repeatedly in the pleural cavity or thigh muscle of rats. In contrast with the effects of hexavalent chromium, no tumors of any type occurred in the pleural cavity, and only one tumor, a sarcoma, developed in the muscle.

The observations on humans referred to two studies in which men were exposed to trivalent chromium by inhalation. In one investigation,[257] acute pneumonitis and fibrosis of the lungs were described in four men employed in a ferroalloy plant where chromite ore was used. The dust in this plant contained many other chemicals, including 36.7% silica (3.35% as quartz, and 1.7% as amorphous silica). However, workers in another, similar ferroalloy plant did not have any such pulmonary conditions.[250] The documentation also quotes an observation in India of exaggerated pulmonic markings in workers exposed to chromite dust.[119]

The only other papers on human exposure[108,224] dealt with allergic dermatitis from patch tests and skin contact with trivalent chromium, but yielded no data for a TLV.

4. *Critique of TLV's for Chromium* The Panel does not believe that the data serving as a basis for the TLV for chromic acid and chromates (hexavalent chromium) are adequate to support the conclusions published in the documentation—namely, that a concentration of chromic acid mists and soluble chromate dust close to 0.1 mg/m^3 (as CrO_3) will prevent nasal irritation and injury and that higher concentrations of sparingly soluble chromates probably are necessary to produce lung cancer. The Panel believes that no sound data were presented in the documentation on which to base a hexavalent-chromium standard for the prevention of lung cancer and that the standard now in common use for prevention of nasal effects requires verification.

The material cited in the documentation contains no data on which to base a TLV for trivalent chromium. The acute pneumonitis and fibrosis found in a ferroalloy plant that used chromite ore were not confirmed in another plant. It seems highly unlikely that chromite dust causes pulmonary disease, inasmuch as it is biologically a very inert acid and a water-insoluble material. Furthermore, the Public Health Service reported very low mortality and lung-cancer rates in men exposed to high concentrations of chromite dust in a chrome brick plant.[97]

ANSI STANDARDS

The American national standard[9] for the allowable concentration of chromic acid and chromates, 0.1 mg/m^3 (as CrO_3), was based on the report of Bloomfield and Blum.[39] The purpose of this standard is "to prescribe the maximum permissible concentration of chromic acid mist or dust from chromates or dichromates of the alkali or alkaline earth metals in the atmosphere of work places for guidance in establishing control procedures for the protection of the health of workers." The standard applies to all places of employment and to exposures not exceeding 8 hr/day. This standard does not appear to include lung-cancer prevention; the only toxic properties listed are dermatitis, ulcers, and perforation of the nasal septum.

OSHA STANDARDS

The standard of 0.1 mg/m^3 (as CrO_3) adopted under the Occupational Safety and Health Act is based on the ACGIH and ANSI values, without reference to the source of the values or to the specific condition to which the standard is applicable.

AMERICAN INDUSTRIAL HYGIENE ASSOCIATION GUIDES

In addition to the standards of the ACGIH, ANSI, and OSHA, the hygiene guides[8] of the American Industrial Hygiene Association (AIHA) should be mentioned. The guides are not specified as standards and have no official status; they are recommendations suggested by AIHA as maximal atmospheric concentrations for an 8-hr workday in industry. The concentration of chromic acid, 0.1 mg/m^3 (as CrO_3), is based on the study by Bloomfield and Blum[39] and another study, which does not give any concentrations. The guide states that the "short exposure tolerance" and the "atmospheric concentration immediately hazardous to life" are unknown and that there is no biochemical assay for exposure. The guide applies only to chromic acid and to the prevention of ulcers and perforation of the nasal septum.

U.S.S.R.

The maximal permissible concentration of chromic anhydride, chromates, and bichromates in the air of workplaces is 0.1 mg/m^3 (as CrO_3). This limit was approved in December 1960.[371 (pp.208,212)] Workplaces are defined as "places at which workers are present regularly or periodically for observing or for carrying out industrial processes. . . . If the industrial operations are carried out at different places in the workroom, the entire workroom is considered a work place. . . . When workers are present in workrooms for short periods, deviations from these figures are permitted with the permission of the State Sanitary Inspection of the USSR."

Italy

The Laboratory for Toxicology of the Clinica del Lavoro in Milan conducted some studies in industries during the early 1950's to determine the validity of the American TLV's for a number of toxic chemicals, including chromates and chromic acid. In a chromate-producing industry in which the concentration of chromates varied from 0.11 to 0.15

mg/m^3, the health effects observed in the workers included nasal septal ulcers, inflammation of the mucous membrane of the larynx and eyes, and chronic bronchitis. Over an observation period of 3 years, one case of nasal septal cancer and one of lung cancer developed in a group of 150 workers. On the basis of this evidence, Vigliani and Zurlo[354] recommended that "the permissible limit be lowered to 0.05 mg/m^3 (as CrO_3) which corresponds to an average excretion through the urine of less than 0.05 mg/liter."

Czechoslovakia

The Czechoslovak Committee of MAC, with Professor Teisinger as chairman,[74] in June 1969 suggested maximal allowable concentrations (MAC's) for chromium chemicals, without distinction between trivalent and hexavalent chromium, as follows:

mean MAC	0.05 mg/m^3, recalculated to Cr
peak MAC	0.1 mg/m^3, recalculated to Cr
mean MAC	0.1 mg/m^3, recalculated to CrO_3
peak MAC	0.2 mg/m^3, recalculated to CrO_3

These values were based on the publications of Bloomfield and Blum[39] and of Vigliani and Zurlo.[354] In addition, the Committee cited the studies of Kubík[175] and Kemka:[74] "Kubík and Kemka found already at mean concentrations 0.049 and 0.077 mg/m^3 Cr O_3 at two different work places, lighter damage of nasal mucous membrane in 54% and 62% of the workers. In another plant, at mean concentrations of 0.432 mg/m^3 they found approx. 50% of lighter damage of nasal mucous membrane and 12% bronchitis and 18% perforations."

The Czechoslovak Committee listed MAC's for various countries and gave the dates when they were reported to the Committee, as follows (for chromium and its salts, calculated as CrO_3):

Federal Republic of Germany (West Germany)	0.1 mg/m^3, 1968
German Democratic Republic (East Germany)	0.1 mg/m^3, 1963
Great Britain	0.1 mg/m^3, 1955
Hungary	0.1 mg/m^3, 1956
Poland	0.1 mg/m^3, 1959
United States	0.1 mg/m^3, 1967
U.S.S.R.	0.1 mg/m^3, 1965
Yugoslavia	0.1 mg/m^3, 1957

Japan

The Japanese Association of Industrial Health[161] has published a list of maximal permissible concentrations of toxic substances for the work environment. These are specified as "time-weighted average concentrations for about 8-hour normal workday." The value recommended in 1969 for chromates was 0.1 mg/m^3 (as CrO_3).

AMBIENT AIR

United States

No standards for ambient-air chromium have been proposed by either the federal government or the individual states.

U.S.S.R.

The criteria used by Russian scientists for the determination of air pollution standards generally differ greatly from those used in the United States. To permit an understanding of the Russian limits, their philosophy and methods are described:[266]

All concentrations of noxious substances that have an unfavourable effect on vegetation, local climate, clearness of the atmosphere and people's living conditions have to be taken into account. It was this standpoint that was taken in the USSR as the basis for working out maximum permissible concentrations. It does not follow from what has been said that legislation on clean air should not take into account considerations of economy and technical attainment. Hygienic standards for maximum permissible concentrations of air pollutants must in themselves reflect the scientifically based ideal towards which we must strive in order to ensure that the public is not subjected to unfavourable effects from air pollution. This ideal cannot be achieved always and everywhere at a given time. Therefore, alongside the general hygienic standards for maximum permissible concentrations, there may be sanitary standards of a temporary character, serving the needs of the moment. They may modify for a defined period the requirements for cleanliness of the external atmosphere, taking into account economic and technological factors. These standards may differ from country to country, as they have to take into account the internal economic situation. Such air pollution standards are permissible temporarily, but should be abandoned after a certain period, during which the condition of the air must be brought into conformity with the hygienic standards. If this approach is adopted, hygienic standards for the cleanness of the external air will not be used to sanction existing technical achievement, but will represent the goal towards which we must strive.

Only in the light of this reasoning can we understand the maximum permissible concentrations of air pollutants adopted in the USSR. In a number of cases these still remain to be achieved, but they represent the ultimate goal and enable us to assess, in each individual case, how far we have advanced in this difficult task. . . .

* * *

Thus, there may be three types of standards limiting the discharge of harmful substances into the air: hygienic, sanitary and technological.

The first meet the requirements of man as a biological species. Their purpose is to ensure such conditions in the external environment as will be fully favourable to man, physiologically speaking, have no ill effects on him and not disturb his well-being.

Sanitary standards are the hygienic standards modified to meet the factors of the immediate situation—the technical state of development and economic possibilities. They constitute a concession to practical material possibilities and are therefore of a temporary nature. They may differ in different countries and in different periods of social and economic development.

The third set of standards—the technological—aims at preventing the discharge of valuable substances into the atmosphere in so far as this is justified by economic considerations.

"Hygienic" standards are based on both animal and human volunteer experiments. Animals are exposed to the toxic chemical for 6 hr/day for several months. The chief emphasis is on chronic changes in the higher nervous activity of the animals, such as conditioned reflexes, chronaxy, and adaptive mechanisms. Cholinesterase activity, coproporphyrin excretion, albumin:globulin ratio, and other characteristics may be measured. In the human experiments, olfactory threshold, nasal irritation, optical chronaxy, respiratory and visual reflexes, adaptation to darkness, electrocortical conditioned reflex, rhythm assimilation of the cerebral cortex, and other measures of cerebral functional activity are used.

These tests were used to determine the "hygienic" standard for hexavalent chromium, with the following results:[177] In human volunteers, the threshold concentration for nasal irritation, reflex activity, and alteration in the functional state of the cerebral cortex was 2.5 $\mu g/m^3$. A concentration of 1.5 $\mu g/m^3$ was without effect. Chronic exposure of rats to 1.5 $\mu g/m^3$ also was without effect, but 30 $\mu g/m^3$ affected the motor chronaxy and carbonic anhydrase in the blood and produced pathologic changes in the respiratory tract. On the basis of these results, it was recommended that the maximal single and 24-hr average concentration of hexavalent chromium be set at 1.5 $\mu g/m^3$ (as CrO_3).

"Sanitary" standards for trivalent chromium and its inorganic compounds (as Cr_2O_3) in the U.S.S.R. were quoted by Katz[163] as 0.25 mg/m^3 for a single measurement and 0.08 mg/m^3 for a 24-hr average.

The Russian standards are difficult to evaluate, because their methods differ greatly from those used in this country. Furthermore, the basic data are not generally available for study.

INDUSTRIAL EMISSION

United States

No United States emission standards are known to the Panel. The concentration of chromium in the ambient air in the immediate vicinity of chromate-producing plants has been reported in only a few publications. In the United States, Bourne and Rushin[41] reported the following concentrations downwind from an old chromate plant during the period June–December 1949:

distance from plant, ft	chromium concentration, mg/m^3 (as Cr)
100	0.12
500	0.1
1,100	0.08
2,000	0.06
10,000	0.001

U.S.S.R.

Russian scientists[177] have recommended some "sanitary clearance zones" around chromium-processing plants related to the amount of chromium emitted by the plant. The highest concentration of hexavalent chromium in the air in the immediate vicinity of two chromate-producing plants was found within a radius of 0.5 km from the plants and amounted to 15–450 $\mu g/m^3$. The concentration did not fall below the recommended maximal allowable concentration (0.0015 $\mu g/m^3$) until a distance of 2 km from the plant.

On the basis of these observations, the "sanitary clearance zones" recommended for chromium-processing plants were: not less than 1,000 m for plants that discharge 200 kg of hexavalent chromium per day and 2,000 m for plants that discharge 1,000 kg/day.

Japan

No official emission standards for Japan are available to the Panel, but some local standards for the discharge of chromium from chromium plants appeared in the "Agreement for Environmental Protection" made between the Nippon Kagaku Kogyo KK (Japan Chemical In-

TABLE 11-5 Japanese Standards for Discharge of Chromium into Air[a]

Area	Chromium Concentration, $\mu g/m^3$ (as CrO_3)	Place of Measurement
Tokuyama City Yamaguchi prefecture (agreement between Japan Chemical Industry and city)	1.5	Edge of building site
Tokyo Guide Standard		
Maximum	1,500	Exit of treatment facility
Minimum	1,000	
Wakayama prefecture law	10	Place of destination, depending on wind direction and speed
	30	Edge of building site
Osaka prefecture law	30	Edge of building site
Hyogo prefecture law	30	Edge of building site
Tokushima prefecture law	1.5	Edge of building site

[a]Derived from *Agreement for Environmental Protection Made by Nippon Kagaku Kogyo KK (Japan Chemical Industry) Tokuyama-Plant.*[1]

dustry) Tokuyama plant and the City of Tokuyama in the Yamaguchi prefecture.[1] These standards are given in Table 11-5.

The only study of the health of inhabitants in the vicinity of a chromate-producing plant was by Kondo *et al.*[170] in the Tachibana area of Japan. The atmospheric concentration of chromic acid along the boundary of the plant premises was below 1.5 $\mu g/m^3$. The study reported no difference between area residents and control-area persons in urinary concentration of chromium or in results of medical examination of blood and urine. No cases of medium or serious sickness were found, although some persons had "light" cases of chronic pharyngitis, which did not require treatment. Among the elementary- and high-school students, there was no increase in complaints or findings indicative of pharyngitis and laryngitis. The authors concluded that there was no definite health injury due to environmental pollution by chromic acid in the area of this plant. The Panel has not found any other studies that relate current concentrations of chromium in ambient air to health.

GENERAL COMMENTS ON STANDARDS FOR CHROMIUM CONTENT OF AIR

This review of current standards for chromium demonstrates the great difficulty in determining safe and harmful concentrations for man when sufficient data are lacking. In the case of lung cancer, the most important effect of chromium in the environment, the evidence of an etiologic relation rests chiefly on cases that developed in the old chromate plants. Before 1950, high concentrations of hexavalent and trivalent and soluble and insoluble dust and mists were always present in the plants, and massive exposures occurred periodically. Few air analyses were made; hence, the type of compound and the concentration responsible for the cancers were not known. The long latent period between initial exposure and development of diagnosable cancers (average, about 20 years) also makes it impossible to determine the dose to which the workers were exposed. Because of reconstruction of chromate plants, which began in about 1950 in this country, the original exposure of older men who develop cancer now cannot be estimated.

In addition to the lack of adequate exposure data, the rates for lung cancer in the old plants were not adjusted for cigarette smoking, the importance of which was not recognized when the early studies were made. In lung-cancer studies today, the role of other chemicals–such as nickel, arsenic, and sulfuric acid–and the role of respiratory viral infections are being considered. The importance of such cofactors in chromate carcinogenesis is unknown. Thus, it is impossible to formulate any sound industrial standards on the basis of these data. It is doubtful whether adequate dose–response relations will ever become available, inasmuch as new automated processes are being developed to prevent exposure of workmen. One new plant has been in operation for 11 years without any evidence of nasal septal perforation or lung cancer.

Standards for air pollution are even more difficult to determine because of the total absence of adequate studies of the general population in relation to chromium in the ambient air. Considering the inability to determine safe and harmful exposures in industry, it is obviously impossible to determine standards for air pollution on the basis of industrial data.

WATER

United States

The maximal allowable concentrations of chromium in drinking water and in wastes discharged into public waterways are set by the individual states. However, they are usually similar to those adopted as standards by the Public Health Service.

The current drinking-water standard, set by the Public Health Service in 1962,[342] states that the presence of hexavalent chromium in excess of 0.05 mg/liter shall constitute grounds for rejection of the supply.

Under the Water Quality Act of 1965, hexavalent chromium may be present only at 5 ppm or less in waters turned into municipal sewage systems, because of the several detrimental effects of hexavalent chromium in sewage treatment.[306] However, recent laboratory and field studies conducted by the Public Health Service concluded that, "short of massive slug doses, chromate alone is unlikely to harm the operation of a sound sewage treatment plant."[341 (p 25)] When a 21-kg slug of chromium as chromic acid was introduced rapidly into a municipal sewage-treatment plant, "no significant adverse effects on the plant processes both aerobic and anaerobic were noted."[341]

European Countries

The World Health Organization[368] in 1970 adopted a standard for hexavalent chromium of 0.05 mg/liter of drinking water for European countries. The accompanying statement follows:

> There are certain chemical substances which, if they are present above certain levels of concentration in supplies of drinking-water, are likely to give rise to actual danger to health. . . .
>
> The presence of any of these substances in excess of the concentrations quoted should constitute grounds for the rejection of the water for use as a piped supply.

The method recommended in the WHO report for chromium was the atomic-absorption spectrophotometric method, which measures total chromium.

U.S.S.R.

In the U.S.S.R., the maximal permissible concentrations adopted in 1960 for water sources were 0.5 mg/liter for trivalent and 0.1 mg/liter for hexavalent chromium.[371]

Japan

The standard suggested by the Environmental Protection Committee in 1970 for the water supply for public use in the Tokyo area was "less than 0.05 ppm" (0.05 mg/liter) for hexavalent chromium.[95]

Conclusions

A hexavalent-chromium concentration of 0.05 mg/liter of drinking water has been the maximal allowable concentration in the United States and other countries for some years. No harmful effects on human health have been reported at this concentration. Whether higher concentrations could be tolerated without ill effect is not known. The NAS–NAE Committee on Water Quality Criteria reviewed the data in 1972 and made the following comments and recommendations for chromium in raw waters to be used for designated purposes:[228]

Drinking Water

The trivalent form is not likely to be present in waters of pH 5 or above because of the very low solubility of the hydrated oxide.

At present, the levels of chromate ion that can be tolerated by man for a lifetime without adverse effects on health are undetermined. It is not known whether cancer will result from ingestion of chromium in any of its valence forms. A family of four individuals is reported to have drunk water for a period of three years with as high as 0.45 mg/l chromium in the hexavalent form without known effects on their health, as determined by a single medical examination (Davids and Lieber 1951).[77*]

Because of adverse physiological effects, and because there are insufficient data on the effect of the defined treatment process on the removal of chromium in the chromate form, it is recommended that public water supply sources for drinking water contain no more than 0.05 mg/l total chromium.

Livestock Waters

Even in its most soluble forms, the element is not readily absorbed by animals, being largely excreted in the feces; and it does not appear to concentrate in any particular mammalian tissue or to increase in these tissues with age (Mertz 1967,[209] Underwood 1971[351]).

An upper allowable limit of 1.0 mg/l for livestock drinking waters is recommended. This provides a suitable margin of safety.

Irrigation Water

Because little is known about the accumulation of chromium in soils in relation to its toxicity, a concentration of less than 1.0 mg/l in irrigation waters is desirable. At this concentration, using 3 acre feet water/acre/yr, more than 80 lb of chromium would be added per acre in 100 years, and using a concentration of 1.0 mg/l for a period of 20 years and applying water at the same rate, about 160 pounds of chromium would be added to the soil in 20 years.

In view of the lack of knowledge concerning chromium accumulation and toxicity, a maximum concentration of 0.1 mg/l is recommended for continuous use on all soils and 1.0 mg/l on neutral and alkaline fine textured soils for a 20-year period are recommended.

*Reference numbers in this extract have been changed to conform with this volume.

12

General Summary and Conclusions

GENERAL SUMMARY

Chromium is found in great abundance in the earth's crust, ranking fourth of the 29 elements of biologic importance. It is present in air and water and in almost all plants and animals. Chromium exists in valences from –2 to +6, but in the environment only Cr^{+3} and Cr^{+6} are of significance. Because of the unreliability and lack of sensitivity of the methods used in sampling and analyzing for chromium, a great deal of erroneous information has appeared in the world literature on environmental concentrations and distribution of chromium. Only in very recent years have better techniques been developed, but at present no analytic methods are acceptable that combine precision, sensitivity, and accuracy with low cost.

According to the most recent data, the concentration of chromium in the ambient air of large industrial cities of the United States usually ranges from 0.01 to 0.03 $\mu g/m^3$ and in nonurban areas is less than 0.01 $\mu g/m^3$. Chromium in the ambient air comes from natural sources, industrial and product uses, and the burning of coal and wood. It is no longer mined in the United States.

The concentration of chromium in soil usually varies from traces to 250 ppm (as chromic oxide), although concentrations as high as 3,900 ppm (and in one case 5%) have been reported in the literature. The concentration in seawater is well below 0.1 ppb, in river water usually between 1 and 10 ppb, and in municipal drinking water from nondetectable to 35 ppb. The amount of chromium in plants varies greatly, depending on the character of the soil. Common foods yield mean chromium concentrations of 0.02–0.22 μg/g.

The total chromium intake per day by man has been estimated as 5–115 μg in food and water and 0.04–0.08 μg in air. In man, neither toxic symptoms as a consequence of excessive dietary intake of chromium nor harmful effects from the chromium in the ambient air are known.

Chromium is an essential trace element, active in very small concentrations. In the hexavalent state, it is a strong oxidizing agent and is able to penetrate biologic membranes. All known biologic interactions result in reduction to the trivalent form and later coordination with organic molecules of living matter, such as proteins and nucleic acids. Deficiency of chromium results in impaired glucose metabolism due to poor effectiveness of insulin. Chromium-responsive forms of impaired glucose tolerance are known to exist in the United States, and chromium deficiency has been demonstrated in malnourished infants in countries in which the diets are low in animal protein. The normal intake of chromium by man is considered marginal at best and is of concern to nutritionists.

Chromium is considered essential for plant growth. In soils deficient in chromium, the addition of this element can stimulate growth. However, large amounts of chromium in the soil may be harmful. No data exist at present to indicate that the normal chromium content of the ambient air exerts any direct effect on the growth or yield of vegetation. Because of the low concentration of chromium in the air, compared with that naturally occurring in the soil and water, the contribution of chromium in the air to the soil and water, and thus to vegetation, is regarded as insignificant.

Almost no data exist concerning the ecologic cycling of chromium in the environment. Data on the chemical valence of chromium in the ecosystem and the environment are not available.

Aquatic species vary greatly in sensitivity to chromium. The concentration of chromium that has been recommended by the NAS–NAE Committee on Water Quality Criteria as safe for almost all species, regardless of valence state, is 0.05 mg/liter.

The known harmful effects of chromium in man are attributable primarily to the hexavalent form. Trivalent chromium is far less toxic than the hexavalent compounds. It is neither irritating nor corrosive. Current knowledge concerning reactions to chromium in man has been obtained almost entirely from experience in occupational exposures and deals chiefly with effects on the skin and respiratory tract.

Cutaneous injury from chromium has been known since 1827. Chromium reactions in the skin are generally classified as primary irritant effects—which include corrosive ulcers, scars, and nonulcerative contact dermatitis—and allergic effects, such as eczematous and noneczematous contact dermatitis. Chromium compounds best known for their ulcerogenic action are strong concentrations of chromic acid, sodium and potassium chromate and dichromate, and ammonium dichromate. Less concentrated hexavalent compounds cause allergic contact dermatitis. It is now known that hexavalent chromium is reduced to the trivalent state within the skin by methionine, cystine, and cysteine. It has been suggested, but not conclusively proved, that the trivalent form resulting from the reduction may form haptene–protein complexes and thereby initiate sensitization. Some investigators believe that all people sensitized to chromium in the hexavalent form are also sensitive to the trivalent form.

Hexavalent chromium is irritating to the respiratory tract and produces ulceration and perforation of the nasal septum. Of far greater significance is the role of chromium in the production of lung cancer. A high incidence of bronchogenic carcinoma has been found in men engaged in the manufacture of chromium chemicals, especially in the old German and United States plants, where the concentration of chromium was extremely high during some phases of the operation. The specific chromium chemicals responsible for these cancers have not been determined. The cancers do not differ from cancer of the lung due to other causes. The latent period between first exposure and onset of disease is usually between 15 and 20 years, but it may be as long as 30 years, or even more. Because of the long latent period and the lack of information on the concentration of chromium in the work environment in past years, no dose–response relation can be stated. Furthermore, data on cigarette smoking were not available. The possible role of cofactors (other chemicals, viruses, etc.) in the production of lung cancer in chromate workers is completely unknown.

In spite of the corrosive action of hexavalent chromium on the skin and nasal septum, cancers of these organs have not been observed. The incidence of cancer of other tissues does not appear to be increased. An excessive risk of lung cancer has not been reported in the industries that use chromium, but this may be because the incidence of lung cancer in the chromium-using industries has not been evaluated.

Acute systemic poisoning due to chromium usually has resulted from accidental exposures, the therapeutic uses of chromium, or suicidal attempts. The principal damage is tubular necrosis of the kidney.

Animal experiments have not been very profitable. Although sarcomas have been produced by injection of some chromium chemicals, their relevance to bronchogenic carcinoma is not clear. However, a few squamous-cell carcinomas of the lungs in rats have resulted from bronchial implantation of pellets or intrapleural injection of calcium chromate or chromate ore roast in media that maintained intimate contact of the chemical with the tissue over a long period. No carcinomas resulted from trivalent chromic oxide. Dose–response evaluation cannot be obtained from these data.

The Panel believes that the data available at present are not adequate for the establishment of any standards for ambient-air concentrations or emissions. No harmful effects of chromium in the ambient air have been reported. The data available from industry are limited and not applicable to air-pollution situations. No data are available to determine the safe concentration of chromium in industrial exposures for prevention of lung cancer. An exposure of 0.1 mg/m^3 for chromic acid or chromates (as CrO_3) is the generally accepted limit in the United States and other countries for prevention of ulceration and perforation of the nasal septum in occupational exposures of 8 hr/day. This limit was based on a 1928 study of chromium platers. The standard requires verification, but the fact that it has been in use for many years without serious results suggests that it has some validity. Because trivalent chromium is much less toxic than hexavalent chromium, standards for the former cannot be deduced from studies on the latter. Although some industrial standards have been proposed for trivalent chromium, basic supporting data are lacking.

The current hexavalent-chromium standard for drinking water, 0.05 mg/liter, has been in use for some years and appears satisfactory. Data are not available to determine whether higher concentrations would be equally safe, as suggested by some observations on animals and humans.

GENERAL CONCLUSIONS

No harmful effects on the health of man or animals are known to have resulted from the presence of chromium in the ambient air or in public drinking water at the current concentrations. Relatively little is known about the influence of chromium on ecologic systems. Because some chromium compounds have caused serious health problems in men through industrial exposure and because the hexavalent compounds are irritating and corrosive to tissues and are highly oxidative, the role of chromium in the environment requires careful consideration.

Evidence that chromium is an essential element for man and animals is accumulating. Chromium-deficiency states have been found in some population groups in the United States and abroad. Therefore, the environmental aspects of chromium should be considered and studied, not only with a view to potential overexposure, but also with regard to deficiency.

13

Recommendations for Future Research

In its examination of the literature on chromium, the Panel found many subjects on which the available data are insufficient to provide definite conclusions. The following recommendations for research, if implemented, will enable rational decisions on the hazard of chromium to be made. These recommendations are not listed in any order of priority.

1. At present, only two analytic techniques can be successfully used for accurate quantitative determination of chromium at the low concentrations that exist in many environmental media, especially in plant and animal tissue—neutron activation and shielded-arc emission spectrography. Both methods are expensive and time-consuming and require considerable experience and thus are not applicable to large-scale environmental studies. Laboratory research, using the latest analytic instrumentation, is needed for the development of sensitive, accurate, and precise methods for the analysis of chromium that could be used by most laboratory investigators. Careful attention must be given to the use of reference methods to ensure proper development of an adequate procedure. In view of recent developments and the availability of relatively inexpensive atomic-absorption spectrophotometers, continued research and development should be pursued in an attempt to devise a satisfactory method for the analysis of chromium in trace amounts. This is particularly appropriate with respect to the recent addition of the heated graphite tube that is now commercially available. Currently, it is recommended that all such research on analytic methods be com-

pared by duplicate analyses with either neutron activation or shielded-arc emission spectrography—preferably using both methods for cross comparison.

2. In environmental studies relating epidemiologic data to the concentration of chromium in various media—tissue, air, water, soil, food products, etc.—it is mandatory that appropriately designed methods of sample collection be used. This is necessary so that the data reported accurately reflect the sample universe from which the subsample is taken. Unfortunately, little is known about the distribution and homogeneity of chromium in various media or about the best method for collection of the sample. These factors, coupled with the need for the design of a statistically valid method of selection of sampling sites and number of samples needed to predict trends accurately, require research studies.

3. Accurate background information on "normal" concentrations of chromium in various media is necessary for predicting trends.

4. The potential toxicity of chromium depends on its valence state. There are no techniques for estimating the concentration of chromium in relation to its valence state, especially in animal and plant tissue. Data of this type also would be extremely useful for understanding the biologic function and availability of chromium. Gas chromatography and extraction by appropriate chelation may be useful in differentiation. Research is required to find satisfactory methods.

5. The chemical interactions of chromium in the environment require study. Recently, attention has been drawn to the possibility of conversion of some inorganic metals to the more volatile and toxic organic state, e.g., methylization under anaerobic and aerobic conditions in sludge.

6. Carefully designed laboratory studies are needed to determine the uptake, distribution, and interactions of various chromium species in ecosystems.

7. Research is needed to ascertain the relation between exposure to airborne chromium and chromium concentrations in urine, blood, and other biologic media, such as hair. If any relation is demonstrated, biologic standards for exposure may become possible.

8. The concentration and type of chromium and duration of exposure responsible for the high incidence of lung cancer in the chromate chemical workers are not known. Long-range studies of the incidence of cancer, adequate analysis of the type and concentration of chromium in the work environment, and careful records of exposure are necessary. It is believed that lung cancer may develop some time after the end of exposure; hence, the data should include information on

persons who have left the industry. Smoking histories must be included.

9. Similar studies should be made to determine the incidence of lung cancer in the chromate-using industries, as distinct from the chromate-producing industry.

10. The generally accepted standard for chromium in air—0.1 mg/m^3 (as CrO_3)—for prevention of nasal injury should be verified in the chrome-plating, anodizing, and other industries in which chromic acid mist is a hazard. Similarly, observations should be made where there is industrial exposure to other hexavalent-chromium chemicals.

11. Data should be collected to set a realistic safe standard for industrial exposure to trivalent chromium. Because of the low toxicity, such standards cannot be deduced from studies in which hexavalent chromium was also present.

12. Although the basic need is to prevent exposure that may be responsible for bronchogenic carcinoma in chromate workers, it is also important to develop methods for the diagnosis of lung cancer early enough for medical and surgical procedures to be useful. Research in cytologic, immunologic, and other biologic techniques directed to this purpose should be promoted.

13. The concentration of chromium in the air around chromium plants and around plants or operations that use chromium should be measured to determine the effectiveness of current air-pollution control methods and for the development of better methods to prevent environmental contamination.

14. Prospective studies of the incidence of lung cancer in population groups in relation to chromium concentration in the ambient air are desirable. However, such studies would be extremely difficult and would require consultation with scientists representing various disciplines early in the planning stage. The data must cover a 20-year period and include smoking, occupational histories, and other variables of the population. The chromium content of the ambient air must be analyzed by valence state, by solubility, and by chemical form, because only specific, slowly soluble, hexavalent compounds that could reach the bronchi have been demonstrated to have carcinogenic properties in animal experiments.

15. Cancer experiments in animals should be continued to gain more information on the specific chemicals that can produce bronchogenic carcinomas, on the mechanism involved, and on methods of prevention. These experiments should include studies on combined exposure—i.e., with other chemicals commonly present in the air of chromate plants and with respiratory viruses. Chronic low-grade-toxicity studies also might be valuable, particularly with regard to kidney damage.

Appendix: Analysis for Chromium

GENERAL REMARKS

The subject of sampling and analytic methods, of particular importance in dealing with chromium, contains many unknowns. The greatest potential error lies in the methods of sampling. The overall environmental program under consideration must be clearly defined before an experimental design is formulated and the investigation initiated. After the accumulation of appropriate field samples, regardless of the type of matrix involved, an analytic method may be appropriately selected on the basis of its capabilities with regard to precision, sensitivity, and accuracy. Chromium presents a serious problem, in that the analytic methods presently acceptable in terms of precision, sensitivity, and accuracy are expensive. Much headway has been made in the last 2 years in analytic development, but these efforts must be strengthened.

Sampling

The selection of sampling methods for air, water, soil, food, vegetation, tissue, etc., is critical and open to considerable discussion. Sampling methods, in general, require a separate review and will not be described in detail here.

Dusts and fumes of chromium compounds may be collected by any method suitable for collection of other dusts and fumes; impingers, electrostatic precipitators, and filters are commonly used.[327] The National Air Surveillance Networks use a high-volume filtration sampler (R. J. Thompson, G. B. Morgan, and L. J. Purdue, personal communication). Chromic acid mists may be collected in an impinger using water or caustic solutions.[362]

In a study of chromium in municipal drinking waters, samples were taken from taps at various places in the city, and the chromium concentrations were averaged. The results may have been affected by possible contamination from chromium in the piping system between water treatment plant and water tap.[346]

It is more difficult to obtain representative samples of tissues. They can be safely collected by dissection with stainless-steel knives or scissors. Tissues are sometimes ground or homogenized to obtain a homogenous mixture from which an aliquot can be taken for analysis. If metal tools are used, contamination from chromium used in the tools themselves is likely because of pressure and heat generated in the process. That applies also to the collection of plant and soil samples. Grinding with an agate mortar and pestle is one of the safest procedures.[58]

Care and uniformity are very important in collecting and preparing samples. Possible reasons for disagreement among chromium concentrations in the literature are the use of nonuniform or unrepresentative samples, contamination in sample preparation, loss in sample preparation due to improper ashing methods, and lack of correction for matrix or interference effects.

Analytic Sensitivity

There are many methods available for detecting chromium. Many have been evaluated by Beyermann.[30,31]

Colorimetric methods involve the formation of colored complexes with reagents like diphenylcarbazide. The diphenylcarbazide complex can be used to detect as little as 3.5 ng of chromium. The relative standard deviation of the method is approximately 3% for 100 ng. Although fairly sensitive, the method is subject to interference by other ions,[30,31] which may account for some of the discrepancies among chromium concentrations reported in the literature.

Flame photometry has been used for the detection of chromium, but this method is less sensitive (by a factor of 10 in some cases) than others, even if special organic solvents are used.[30,31] A lower detection limit (5 ppb), using this method, has been reported by some investigators.[248]

Emission spectrography, a very specific method, is sensitive to 0.1 ng, particularly when used with modern modifications.[118] X-ray emission spectrography is not quite sensitive enough for use with biologic samples, unless preceded by considerable concentration.[30,31]

Polarography has been used to study the interaction between chromium, insulin, and mitochondrial membranes.[59] However, it is not sensitive enough for detecting chromium in biologic material, because its limit of detection is approximately 50 ng.[210] A rapid polarographic technique has been reported[48] for measuring several metals, including chromium, in sewage and industrial wastes at concentrations of 5 ppm or greater. Other polarographic methods reported in the literature are sensitive to 1 ng, but these are subject to considerable interference by other ions.[30,31] Neutron-activation analysis has been used for detection of chromium in biologic material[64] and in air filters. The detection limit depends on the duration of irradiation and the thermal-neutron flux: the longer the irradiation and the higher the flux, the lower the detection limit. A 5-hr irradiation at 1.5×10^{13} neutrons/cm^2-sec gives a detection limit of 20 ng for air samples analyzed nondestructively. However, neutron-activation analysis requires very expensive equipment, long irradiation times, and highly trained personnel. Separation should improve the detection limit by reducing the background radiation. Gas–liquid chromatography of metal complexes of hexafluoroacetylacetone or related compounds appears to be a very promising new tool, with a sensitivity to 0.01 ng.[12,310] A method using trifluoroacetylacetone, with a sensitivity of 1 ng/100 ml, has been applied to the detection of chromium in serum.[289] At present, the most widely used procedure is atomic-absorption spectrophotometry. The sensitivity of this method can be improved to approximately 6 ng by the use of organic solvents, such as methylisobutylketone, which will also extract hexavalent chromium from its aqueous solution. The method therefore requires digestion and oxidation of the sample.[99,100,210] New developments using graphite-furnace techniques in conjunction with atomic absorption reportedly yield detection limits of 0.05 ng and require only small volumes of solution for analysis, but published data are not available. The concentrations of chromic acid mists in air can be estimated by a direct field method described by Ege and Silverman.[91,311,312] This is a spot-test method that uses phthalic anhydride and s-diphenylcarbazide.

Hexavalent Versus Trivalent Chromium

Because hexavalent chromium is considered more toxic than the trivalent form, it would often be useful to determine the concentration of hexavalent chromium, in addition to the total chromium concentration.

This is possible with all methods that require oxidation to the highest oxidation state for solvent extraction or for formation of a colored complex. Omission of the oxidation steps would result in the determination of only naturally occurring hexavalent chromium. A parallel assay with the oxidation steps included would yield the total chromium content of the sample; the difference between the two would be a measure of the trivalent form.[210]

Precision

The precision of a given method at its detection limit is very poor (100% or greater error). Precision improves with increasing concentration, reaching a maximal error of 1–5% at concentrations about 50–100 times the detection limit. Detection limits are constantly being lowered by advancements in technology. Care must be taken to avoid erroneous results due to interferences, especially near the detection limits. In the diphenylcarbazide method, for example, other ions may form colored complexes with the diphenylcarbazide and have some absorbance at the wavelength of the chromium diphenylcarbazide. In neutron-activation analysis, another element could have a gamma energy of the same energy as chromium, thus masking the peak; this can be checked by looking at the half-life of the 320-keV peak. In gas–liquid chromatography, it is possible for two compounds to have the same retention times on a column; therefore, it is a good idea to use two columns of rather different polarity, so that retention times will change. In atomic-absorption spectrophotometry, at low chromium concentrations, the edge of some other ion's absorption may give a signal that could be mistaken for that of chromium. Proper selection of wavelength will minimize this error.

SPECIFIC ANALYTIC METHODS

Diphenylcarbazide Method

Acid chromate solutions react with diphenylcarbazide to produce a violet solution, with maximal absorption at a wavelength of 540 nm. This color reaction is very sensitive to small concentrations of chromium, with sensitivities of 0.02–0.05 μg/ml;[42] 1 ppm can be detected. Oxidants interfere with the analysis by decomposing the complex, and they must be removed. Vanadium interferes if the vanadium : chromium ratio is larger than 1 : 10; in such a case, vanadium must be removed with oxine (8-hydroxyquinoline).[251]

The sample is subjected to alkaline fusion, using 2.5–5 g of sodium carbonate per 0.25–0.5 g of a sample containing 1–500 μg of chromium. The mixture is then dissolved in a few milliliters of hot water and brought to a boil, with a few drops of ethyl alcohol added to reduce the MnO_4^- ions. The solution is filtered, and the residue is washed with water containing a little carbonate. Sulfuric acid is then added to neutralize the carbonate ions. The acidity of the resulting solution is adjusted to 0.2N. One or two milliliters of diphenylcarbazide in 1 : 1 water–acetone mixture is added. The solution is diluted to 25 or 50 ml; after 10–15 min, the absorption is measured at a wavelength of 540 nm.[251]

If vanadium is present in excessive amounts, it is extracted with oxine. The initial solution is adjusted to a pH of 4.4 with sulfuric acid, with methyl orange as an indicator; 0.1 ml of a 2.5% solution of oxine in 2N acetic acid is added; and the solution is extracted three times with 3 ml of chloroform. After separation, the aqueous phase is freed from chloroform by evaporation and filtered. Sulfuric acid is added to make the solution 0.2N, and 1 ml of diphenylcarbazide is added.[251]

If reduction of the chromates is suspected, 5 drops of nitric acid and 1 ml of 2.5% ammonium persulfate are added to the sample solution, which is boiled for a few minutes and then cooled before the diphenylcarbazide is added.[251]

This method is applicable to all complex media, such as soils, rocks, ores, and plant and animal products.[251,352,353]

Polarographic Techniques

The reduction of hexavalent chromium to the trivalent state in an alkaline medium (sodium hydroxide) yields a well-defined wave at a potential of −0.85 V, with respect to a saturated calomel electrode:

$$CrO_4^{-2} + 4H_2O + 3e^- \rightarrow Cr(OH)_3 + 5OH^-.$$

This is the most characteristic polarographic reaction of chromium, and it is the basis of the determination of chromium in metallic alloys, minerals, soils, and water; it is sensitive to a few micrograms of chromium per milliliter.[251]

A 0.5- to 1-g sample of soil is solubilized after fusion with sodium peroxide, with sufficient hydrochloric acid to yield a 1–2N sodium hydroxide solution. The polarogram is recorded between −0.5 and −1.2 V; the chromium wave is at −0.85 V.

For plant ash, a 0.5-g sample is fused with 2 g of sodium peroxide and redissolved in water and hydrochloric acid to yield a 1–2N sodium hydroxide solution. The polarogram is then recorded between −0.5 and

−1.2 V. For sewage and industrial wastes with initial chromium concentrations above 5 ppm, the sample is agitated to obtain a homogeneous suspension of solids. A 1-liter sample is measured out with a volumetric flask. The sample is acidified with 5 ml of concentrated nitric acid and evaporated to 15–20 ml in a large evaporating dish. The solution is transferred, with any solids remaining in the dish, to 125 ml of 60% perchloric acid. The mixture is evaporated slowly on a hot plate until dense white fumes of perchloric acid appear in the flask, and then fumed for 4–5 min to ensure complete oxidation of chromium. It is cooled somewhat, 50 ml of water is added, and the sample is boiled to expel chlorine. The mixture is then cooled and transferred to a 100-ml volumetric flask. If lead is present, 10 ml of 10% sodium sulfate is added to precipitate the lead as lead sulfate; then the mixture is diluted to the mark with water. A 25-ml aliquot is transferred to a 125-ml conical flask, a drop of phenolphthalein is added, and the sample is neutralized with 2M sodium hydroxide. After 50 ml excess 2M sodium hydroxide is added, the sample is heated to boiling, cooled, transferred to a 100-ml volumetric flask, and diluted to volume with water. Nitrogen is passed through the sample for 10 min, and then the current is measured at −0.60 and −1.10 V versus the saturated calomel electrode.[48]

Neutron-Activation Analysis

Neutron-activation analysis has been used for analysis of chromium in air samples, alloys, and biologic samples. A possible advantage of neutron activation is that it permits nondestructive analysis in many cases. Chromium-50 has a low abundance (4.31%), a small thermal-neutron cross section (17 barns), and a low percentage of gamma decay. The half-life of chromium-51 is 27.8 days. These factors make it necessary to use long irradiations at high flux in order to study low concentrations of chromium.

A portion of the air filter is encapsulated in quartz or heat-sealed in a polyethylene tube and irradiated for 2–5 hr in a flux of 1.5×10^{13} neutrons/cm^2-sec. After 20–30 days, the sample is counted for 4,000 sec, and the 320-keV gamma peak is counted. The detection limit is 0.02 μg.[76] This might be improved by longer irradiation or higher flux.

Biologic samples must be encapsulated in quartz, and the long irradiation will usually char the material. Most often, a separation will be required. A carrier containing 100 mg of chromium should be added, followed by digestion in nitric acid, nitric and perchloric acids, or nitric, sulfuric, and perchloric acids, depending on the matrix. Care should be taken to avoid any addition of chlorine, because chromyl chloride is volatile. After digestion, perchloric acid is added and the solution is evaporated, yielding fumes of perchloric acid. The chromyl chloride can then be distilled by heating the solution as gaseous hydrogen chloride is bubbled through the solution. The yield can be determined by precipitating barium chromate and weighing.

Gas–Liquid Chromatography

Gas–liquid chromatography of chromium involves formation of volatile chelates with acetylacetone, trifluoroacetylacetone (tfa), and hexafluoroacetylacetone (hfa). Procedures for converting several metals in biologic materials to the fluorine-substituted acetylacetones have been described—e.g., for the detection of beryllium in blood, urine, and tissue (these procedures can probably be applied to other metals) and for the detection of chromium in serum. Detection limits as low as 3×10^{-12} g of chromium (3 picograms) have been attained with the hfa complex.[225,274,309] However, many investigators have abandoned the gas–liquid chromatographic method for the determination of chromium, especially in various biologic materials, such as serum. Interfering substances have often led to spurious high values. It appears from unpublished data that additional developmental research is necessary before this technique can be used as a routine tool for the determination of chromium.

A glass reaction tube (4 in. long and 3/16 in. in inside diameter) is made by sealing the tapered end of a disposable pipette with an oxygen–methane torch. A 0.050-ml sample of the biologic material to be tested is introduced into the glass tube, and then 0.050 ml of a hexane solution containing 0.005 ml of tfa. The sealed end is immersed in an ice bath, and the open end of the tube is then sealed at the constriction. The sealed tube is shaken for 15 sec, wrapped in heavy-duty aluminum foil, and placed on its side in a 175 C oven for 30 min. After 5 min of cooling and 5 min of centrifuging, the reaction tube is opened and a 0.40-ml aliquot of the hexane phase is added to 0.50 ml of 1.0N sodium hydroxide in a 2-dram vial, which is then sealed with a polyethylene-lined cap. After 5 min of shaking, the vial is centrifuged for 5 min. A sample of the hexane phase is placed in a small vial, which is then tightly sealed. Five replicated injections of 1.0 μl of both standard solutions and the hexane phase of the samples being analyzed are usually required for precise analyses. Time can be saved by using fewer injections, with some sacrifice in precision. The average *trans*-$Cr(tfa)_3$ peak heights of the standard $Cr(tfa)_3$ solutions are compared with those from the hexane solution from the reaction to calculate the concentration of chromium in the sample. The sample preparation time is about 60 min. The instrument time per injection is 12 min. If four or more simultaneous analyses are performed, the average time is less than 1 hr.[350]

The electron-capture detector is very sensitive to H(tfa), $Fe(tfa)_3$, and other volatile, hexane-soluble species with high electron affinities. The large excess of these materials may mask the Cr(tfa) peaks, cause the detector to lose its sensitivity, or give a noisy, highly sloped baseline. Therefore, it is necessary to back-extract as much of the interfering material from the hexane as possible.[350]

Sievers *et al.*[310] reported that a backwash with 1.0N sodium hydroxide successfully removed excess H(tfa) and $Fe(tfa)_3$ with no loss of $Cr(tfa)_3$. They used 5-min back-extraction with 1.0N sodium hydroxide followed by centrifuging and layer separation.

Atomic-Absorption Spectrophotometry

For routine work, atomic-absorption spectrophotometry appears to be satisfactory in terms of precision and economy. Most samples require ashing and oxidation. In many systems, it is possible to analyze for chromium without extraction into an organic solvent. The detection limit is about 0.01 or 0.02 mg/liter. The general instrument characteristics are: chromium hollow-cathode lamp; wavelength, 3579 Å; boiling burner; fuel, acetylene; oxidant, air; and type of flame, slightly fuel-rich.[350]

New developments in atomic-absorption spectrophotometry are occurring rapidly, including improvements in design and new accessory modules for increasing sensitivity and reducing interferences. Among these are the deuterium background corrector for reducing interferences from light scattering in the flame due to the presence of extraneous cations, such as sodium and potassium, and the "heated-graphite atomizer" for atomic absorption. The latter has great potential, especially for the analysis of chromium, but much exploratory work needs to be done before it becomes available for the routine analysis of chromium. There is almost no published information on the application of this method to the analysis of chromium, although private communications indicate extreme improvements in detection limits. The sample matrix is a crucial variable, which must be carefully controlled. It also appears that the viscosity of the sample contributes to the geometric distribution of the sample within the graphite tube, which affects the observed peak height. At this stage, it would be premature to suggest the use of this device until additional studies are conducted on some obvious variables that affect the analysis of chromium.

For biologic samples, 1–10 g is added to a 30-ml Kjeldahl flask, followed by 10 ml of a 3 : 1 : 1 mixture of nitric, sulfuric, and perchloric acids. Analysis requires at least 3 ml of serum. The mixture is heated gently, at first, to allow the protein to go into solution. The heat is then increased until dense fumes of sulfur trioxide appear, indicating that digestion is complete. The flask is cooled, and 5 ml of distilled water and 1 ml of 0.01M permanganate are added. The solution is boiled gently for 5 min to allow the oxidation of Cr^{+3} to Cr^{+6}. If the pink color disappears, more permanganate is added until the color persists. The solution is cooled to 5 C in an ice bath, and then transferred to a 30-ml separatory funnel. The flask is washed twice with 1-ml portions of distilled water and the water is added to the separatory funnel. Then 2 ml of concentrated hydrochloric acid and 5 ml of methylisobutylketone are added; both were cooled to 5 C before addition. The mixture is shaken for 30 sec, and the ketone layer is separated. The solution is then aspirated into the flame. The extraction step must not be done too slowly, lest the chromium be reduced to the trivalent form, which would not extract.[58]

Samples difficult to wet-ash have been determined by first dry-ashing at 550 C in a porcelain crucible, taking up the ash in 0.5 ml of hydrochloric acid, transferring it to a Kjeldahl flask, and running it through the entire wet-ashing and solvent-extraction procedure. A 5-g sample is generally used.[58,99]

Another recently reported advance in the analysis of plants and other biologic materials for chromium involves a refinement of preconcentration techniques and determination of chromium by atomic-absorption spectrophotometry. The method uses a wet-digestion procedure, isolation of chromium by solvent extraction, and extraction of the chromium chelate of 2,4-pentanedione into chloroform. The chloroform is evaporated, and the chromium is taken up in 4-methyl-2-pentanone.[56] Although this procedure is rather rigid and time-consuming, recoveries were 95%, with very good sensitivity and precision, on the basis of interlaboratory comparisons. The authors report that this method allows for estimation of 4 ng of chromium per milliliter of solution and that interferences are minimal.

Emission Spectroscopy

A recent publication by Hambidge[131] describes an application of arcing in a static argon atmosphere using a refined excitation chamber for the determination of chromium in biologic media. This technique was applied to the trace analysis of ashed biologic material, especially the measurement of nanogram quantities of chromium in blood, hair, and urine. The mean relative standard deviation for quantities of chromium ranging from 1 to 7 ng in 0.2-ml aliquots of serum was 6%. Sample preparation involved ashing by oxidizing the organic material in the oxygen plasma of a low-temperature asher. Preliminary data indicate this method to be potentially excellent, especially as a research tool for analysis of chromium in biologic tissue. The reportedly greater sensitivity of this method than of conventional atomic-absorption spectrophotometry is a significant advantage in the measurement of very small amounts of chromium—in the nanogram range.[131]

References

1. Agreement for environmental protection made by Nippon Kagaku Kogyo KK (Japan Chemical Industry) Tokuyama-Plant. Sangyo Kogai (Industrial Public Nuisance) 6:743–747, 1970. (in Japanese)
2. Ahrens, L. H. Distribution of the Elements in our Planet. New York: McGraw-Hill, Inc., 1965. 110 pp.
3. Akatsuka, K., and L. T. Fairhall. The toxicology of chromium. J. Ind. Hyg. 16:1–24, 1934.
4. Alwens, W. Lungenkrebs durch Arbeit in Chromat herstellenden Betrieben. (b) Klinischer Teil, Vol. 2., pp. 973–982. In Bericht über den VIII. Internationalen Kongress für Unfallmedizin und Berufskrankheiten. Frankfort A.M., 26–28 September 1938. (2 vols.) Leipzig: Thieme, 1939.
5. Alwens, W., and W. Jonas. Der Chromat-Lungenkrebs. Acta Un. Int. Cancr. 3:103–118, 1938.
6. American Conference of Governmental Industrial Hygienists. Documentation of the Threshold Limit Values for Substances in Workroom Air. (3rd ed.) Cincinnati, Ohio: American Conference of Governmental Industrial Hygienists, 1971. 286 pp.
7. American Conference of Governmental Industrial Hygienists. Threshold Limit Values of Airborne Contaminants and Physical Agents with Intended Changes. Adopted by ACGIH for 1971. Cincinnati: American Conference of Governmental Industrial Hygienists, 1971. 82 pp.
8. American Industrial Hygiene Association. Hygienic Guide Series. Chromic acid. Amer. Ind. Hyg. Assoc. J. 17:233–235, 1956.
9. American National Standards Institute, Inc. Allowable Concentration of Chromic Acid and Chromates. USAS Z37.7–1943 (rev. 1971). New York: American National Standards Institute, Inc., 1971. 7 pp.

10. Amiel, J. Chrome, pp. 33–413. In Nouveau Traité de Chimie Minérale. Vol. XIV. Publie sous la direction de P. Pascal. Paris: Masson et Cie, 1959.
11. Anwar, R. A., R. F. Langham, C. A. Hoppert, B. V. Alfredson, and R. U. Byerrum. Chronic toxicity studies. III. Chronic toxicity of cadmium and chromium in dogs. A.M.A. Arch. Environ. Health 3:456–460, 1961.
12. Aue, W. A. Current capabilities in analysis of trace substances: Gas chromatography, pp. 37–44. In D. D. Hemphill, Ed. Proceedings of University of Missouri's 1st Annual Conference on Trace Substances in Environmental Health. Columbia: University of Missouri, 1967.
13. Baader, E. W. Berufskrebs, pp. 104–128. In C. Adam and H. Auler, Eds. Neuere Ergebnisse auf dem Gebiete der Krebskrankheiten. Leipzig: Hirzel, 1937.
14. Bacon, F. E. Chromium, pp. 1200–1201. In T. Lyman, Ed. Metals Handbook. Vol. I. Properties and Selection of Metals. (8th ed.) Metals Park, Ohio: American Society for Metals, 1961.
15. Bacon, F. E. Chromium and chromium alloys, pp. 451–472. In R. E. Kirk and D. F. Othmer, Eds. Kirk–Othmer Encyclopedia of Chemical Technology. Vol. 5. (2nd ed.) New York: Interscience Publishers, 1966.
16. Baetjer, A. M. Pulmonary carcinoma in chromate workers. I. A review of the literature and report of cases. A.M.A. Arch. Hyg. Occup. Med. 2:487–504, 1950.
17. Baetjer, A. M. Pulmonary carcinoma in chromate workers. II. Incidence on basis of hospital records. A.M.A. Arch. Ind. Hyg. Occup. Med. 2:505–516, 1950.
18. Baetjer, A. M. Relation of chromium to health, pp. 76–104. In M. J. Udy, Ed. Chromium. Vol. I. Chemistry of Chromium and Its Compounds. American Chemical Society Monograph #132. New York: Reinhold Publishing Corporation, 1956.
19. Baetjer, A. M., C. Damron, and V. Budacz. The distribution and retention of chromium in men and animals. A.M.A. Arch. Ind. Health 20:136–150, 1959.
20. Baetjer, A. M., J. F. Lowney, H. Steffee, and V. Budacz. Effect of chromium on incidence of lung tumors in mice and rats. A.M.A. Arch. Ind. Health 20:124–135, 1959.
21. Barbera, L. Patologa da acido cromico e derivati. Rass. Med. Appl. Lav. Ind. 6:211–239, 1935.
22. Barrows, H. L. The agricultural significance of inorganic pollutants, pp. 73–88. In Proceedings, Maryland Farm and Land Brokers' Institute and Maryland Agri-Business Resources Seminar, February 20–21, 1968, University of Maryland, College Park. College Park: University of Maryland.
23. Bauer, K. H. Referat über Berufsschaden und Krebs. Verh. Dtsch. Path. Ges. 30:239–286, 1937.
24. Bécourt, M. M., and A. Chevallier. Memoire sur les accidents qui atteignent les ouvriers qui travaillent le bichromate de potasse. Ann. d'Hyg. Publ. Med. Leg. 20:83–95, 1863.
25. Berrow, M. L., and J. Webber. Trace elements in sewage sludge. J. Sci. Food Agric. 23:93–100, 1972.
26. Bertine, K. K., and B. D. Goldberg. Fossil fuel combustion and the major sedimentary cycle. Science 173:233–235, 1971.
27. Bertrand, D. Va-t-il falloir ajouter le chrome aux oligo-éléments devant être utilisés comme engrais complémentaires? C. R. Hebd. Seances Acad. Agr. 53:113–115, 1967.

28. Bertrand, D., and A. deWolf. Le chrome, oligoélément dynamique pour les végétaux supérieurs. C. R. Acad. Sci. 26:5616-5617, 1965.
29. Bertrand, D., and A. deWolf. Nécessité de l'oligo-élément chrome pour la culture de la pomme de terre. C. R. Acad. Sci., Ser. D 266:1494-1495, 1968.
30. Beyermann, K. Das analytische Verhalten kleinster Chrommengen. Teil I. Z. Anal. Chem. 190:4-33, 1962.
31. Beyermann, K. Das analytische Verhalten kleinster Chrommengen. Teil II. Z. Anal. Chem. 191:346-369, 1962.
32. Bidstrup, L. Industrial aspects. Cancer of the lungs, nose and nasal sinuses, pp. 81-97. In R. W. Raven, Ed. Cancer Progress. London: Butterworths, 1960.
33. Bidstrup, P. L. Bronchi and lungs–industrial factors, pp. 193-203. In R. W. Raven and F. J. C. Roe, Eds. The Prevention of Cancer. London: Butterworths, 1967.
34. Bidstrup, P. L. Carcinoma of the lung in chromate workers. Brit. J. Ind. Med. 8:302-305, 1951.
35. Bidstrup, P. L. Chromium, alloys, compounds, pp. 294-297. In Encyclopaedia of Occupational Safety and Health. Vol. 1/A-K. Geneva: International Labour Office, 1971.
36. Bidstrup, P. L., and R. A. M. Case. Carcinoma of the lung in workmen in the bichromates-producing industry in Great Britain. Brit. J. Ind. Med. 13:260-264, 1956.
37. Birmingham, D. J. Skin hygiene and dermatitis in industry. Arch. Environ. Health 10:653-657, 1965.
38. Blair, J. Chrome ulcers. Report of twelve cases. J.A.M.A. 90:1927-1928, 1928.
39. Bloomfield, J. J., and W. Blum. Health hazards in chromium plating. Public Health Rep. 43:2330-2351, 1928.
40. Bockendahl, H. Chromnacheweis und Chromgehalt gefärbter Kleiderstoffe. Ein Beitrag zur Bedeutung gefärbter Kleiderstoffe für die Häufigkeit der Chromallergie. Derm. Wschr. 130:987-991, 1954.
41. Bourne, H. G., Jr., and W. R. Rushin. Atmospheric pollution in the vicinity of a chromate plant. Ind. Med. 19:568-569, 1950.
42. Bowen, H. J. M. Trace Elements in Biochemistry. New York: Academic Press, 1966. 241 pp.
43. Brard, D. Toxicologie du Chrome. Paris: Hermann and Co., 1935. 80 pp.
44. Brieger, H. Zur Klinik der akuten Chromatvergiftung. Z. Exp. Path. Therap. 21:393-408, 1920.
45. Broch, C. Bronchial asthma caused by chromium trioxide fumes. Nord. Med. 41:996-997, 1949. (in Swedish) [Arch. Ind. Hyg. Occup. Med. 1:588, 1950 (abstract).]
46. Buckell, M., and D. G. Harvey. An environmental study of the chromate industry. Brit. J. Ind. Med. 8:298-301, 1951.
47. Burkeholder, J. N., and W. Mertz. Properties and effects of chromium(III) fractions obtained from brewer's yeast, pp. 701-705. In Proceedings of the Seventh International Congress of Nutrition, Hamburg, 1966. Vol. V. Physiology and Biochemistry of Food Components. New York: Pergamon Press, 1967.
48. Butts, P. G., and M. G. Mellon. Industrial wastes. Polarographic determination of metals in industrial wastes. Sew. Ind. Wastes 23:59-63, 1951.
49. Byerrum, R. U. Some studies on the chronic toxicity of cadmium and hexavalent chromium in drinking water, pp. 1-8. In Proceedings of the Fifteenth

Industrial Waste Conference, May 3, 4, and 5, 1960. Engineering Extension Series No. 106. Lafayette, Ind.: Purdue University, 1961.
50. Cairns, J., Jr. The effects of increased temperature upon aquatic organisms. In 10th Industrial Waste Conference Proceedings, Purdue University Eng. Bull. 40:346–354, 1956.
51. Cairns, R. J., and C. D. Calnan. Green tattoo reactions associated with cement dermatitis. Brit. J. Derm. 74:288–294, 1962.
52. Calnan, C. Cement dermatitis. J. Occup. Med. 2:15–22, 1960.
53. Campbell, W. J., and W. Mertz. Interaction of insulin and chromium(III) on mitochondrial swelling. Amer. J. Physiol. 204:1028–1030, 1963.
54. Carter, B. B., D. F. Jackson, and A. R. Kolber. Observations on the attachment of chromium-51 to the human red cell. Int. J. Appl. Radiat. Isot. 18:615–618, 1967.
55. Carter, J. P., A. Kattab, K. Abd-El-Hadi, J. T. Davis, A. El Gholmy, and V. N. Pathwardhan. Chromium(III) in hypoglycemia and in impaired glucose utilization in kwashiorkor. Amer. J. Clin. Nutr. 21:195–202, 1968.
56. Cary, E. E., and W. H. Allaway. Determination of chromium in plants and other biological materials. Agric. Food Chem. 19:1159–1161, 1971.
57. Chargaff, E., and C. Green. On the inhibition of the thromboplastic effect. J. Biol. Chem. 173:263–270, 1948.
58. Christian, G. D., and F. J. Feldman. Atomic Absorption Spectroscopy. Applications in Agriculture, Biology, and Medicine. New York: Wiley-Interscience, 1970. 490 pp.
59. Christian, G. D., E. C. Knoblock, W. C. Purdy, and W. Mertz. A polarographic study of chromium–insulin–mitochondrial interaction. Biochim. Biophys. Acta 66:420–423, 1963.
60. Chuecas, L., and J. P. Riley. The spectrophotometric determination of chromium in sea water. Anal. Chim. Acta 35:240–246, 1966.
61. Clarke, G. D. The effect of cobaltous ions on the formation of toxin and coproporphyrin by a strain of *Corynebacterium diphtheriae.* J. Gen. Microbiol. 18: 708–719, 1958.
62. Clendenning, K. A., and W. J. North. Effects of wastes on the giant kelp, Macrocystis pyrifera, pp. 82–91. In E. A. Pearson, Ed. Proceedings of the First International Conference on Waste Disposal in the Marine Environment, University of California, Berkeley, July 22–25, 1959. New York: Pergamon Press, 1960.
63. Cohen, H. A. Experimental production of circulating antibodies to chromium. J. Invest. Derm. 38:13–20, 1962.
64. Coleman, R. F., F. H. Cripps, A. Stimson, and H. D. Scott. The Determination of Trace Elements in Human Hair by Neutron Activation and the Application to Forensic Science. Atomic Weapons Research Establishment Report AWRE-0-86/66. Aldermaston, Eng.: Atomic Weapons Research Establishment, 1967. 37 pp.
65. Collins, R. J., P. O. Fromm, and W. D. Collings. Chromium excretion in the dog. Amer. J. Physiol. 201:795–798, 1961.
66. Conn, L. W., H. L. Webster, and A. H. Johnson. Chromium toxicology. Absorption of chromium by the rat when milk containing chromium was fed. Amer. J. Hyg. 15:760–765, 1932.
67. Corwin, L. M., G. R. Fanning, F. Feldman, and P. Margolin. Mutation leading to increased sensitivity to chromium in *Salmonella typhimurium.* J. Bact. 91:1509–1515, 1966.

68. Cotton, F. A., and G. Wilkinson. Advanced Inorganic Chemistry. A Comprehensive Text. (2nd ed.) New York: Interscience Publishers, 1966. 1,136 pp.
69. Cuffe, S. T., and R. W. Gerstle. Emissions from Coal-Fired Power Plants: A Comprehensive Summary. U.S. Department of Health, Education, and Welfare, Public Health Service Publication 999-AP-35. Cincinnati: Public Health Service, 1967. 26 pp.
70. Cumin, W. Remarks on the medicinal properties of madar and on the effects of bichromate of potass on the human body. Edinburgh Med. Surg. J. 28: 295–302, 1827.
71. Curl, H., Jr., N. Cutshall, and C. Osterberg. Uptake of chromium(III) by particles in sea-water. Nature 205:275–276, 1965.
72. Curran, G. L. Effect of certain transition group elements on hepatic synthesis of cholesterol in the rat. J. Biol. Chem. 210:765–770, 1954.
73. Cutshall, N., V. Johnson, and C. Osterberg. Chromium-51 in sea water: Chemistry. Science 152:202–203, 1966.
74. Czechoslovak Committee of MAC. Documentation of MAC in Czechoslovakia. Prague: Czechoslovakia Ministry of Health, 1969. 167 pp.
75. DaCosta, J. G., J. F. X. Jones, and R. C. Rosenberger. Tanner's ulcer: Chrome sores–chrome holes–acid bites. A.M.A. Ann. Surg. 63:155–166, 1916.
76. Dams, R., J. A. Robbins, K. A. Rahn, and J. W. Winchester. Nondestructive neutron activation analysis of air pollution particulates. Anal. Chem. 42:861–867, 1970.
77. Davids, H. W., and M. Lieber. Underground water contamination by chromium wastes. Wat. Sew. Works 98:528–534, 1951.
78. Davidson, I. W. F., and W. L. Blackwell. Changes in carbohydrate metabolism of squirrel monkeys with chromium dietary supplementation. Proc. Soc. Exp. Biol. Med. 127:66–70, 1968.
79. Delpech, M. A., and M. Hillairet. Mémoire sur les accidents auxquels sont soumis les ouvriers: Employés a la fabrication des chromates. Ann. d'Hyg. Publ. Med. Leg. 31:5–30, 1869.
80. Denton, C. R., R. G. Keenan, and D. J. Birmingham. The chromium content of cement and its significance in cement dermatitis. J. Invest. Derm. 23:189–192, 1954.
81. Desbaumes, P., and D. Ramaciotti. Étude chimique de l'action sur la végétation d'un effluent gazeux industriel dontenant du chrome hexavalent. Pollut. Atmos. 10:224–226, 1968.
82. Dobrolyubovskii, O. K., and A. V. Slavvo. Application of new trace nutrients containing chromium in grape culture. Udobr. Uroz. 3:35–37, 1958. (in Russian) (Chem. Abstr. 53:631e)
83. Doisy, R. J., D. H. P. Streeten, M. L. Souma, M. E. Kalafer, S. I. Rekant, and T. G. Dalakos. Metabolism of ^{51}chromium in human subjects, normal, elderly, and diabetic subjects, pp. 155–168. In W. Mertz and W. E. Cornatzer, Eds. Newer Trace Elements in Nutrition. New York: Marcel Dekker, Inc., 1971.
84. Doudoroff, P., and M. Katz. Critical review of literature on the toxicity of industrial wastes and their components to fish. II. The metals as salts. Sew. Ind. Wastes 25:802–839, 1953.
85. Ducatel, J. T. On poisoning with preparations of chrome. Baltimore Med. Surg. J. Rev. 1:44–49, 1833.
86. Durfor, C. N., and E. Becker. Public Water Supplies of the 100 Largest Cities

in the United States, 1962. U.S. Geological Survey Water-Supply Paper 1812. Washington, D.C.: U.S. Government Printing Office, 1964. 364 pp.

87. Durum, W. H., and J. Haffty. Implications of the minor element content of some major streams of the world. Geochim. Cosmochim. Acta 27:1–11, 1963.

88. Durum, W. H., J. D. Hem, and S. G. Heidel. Reconnaissance of Selected Minor Elements in Surface Waters of the United States, October 1970. U.S. Geological Survey Circular 643. Washington, D.C.: U.S. Geological Survey, 1971. 49 pp.

89. Edmundson, W. F. Chrome ulcers of the skin and nasal septum and their relation to patch testing. J. Invest. Derm. 17:17–19, 1951.

90. Edwards, C., K. B. Olson, G. Heggen, and J. Glenn. Intracellular distribution of trace elements in liver tissue. Proc. Soc. Exp. Biol. Med. 107:94–97, 1961.

91. Ege, J. F., Jr., and L. Silverman. Stable colorimetric reagent for chromium. Anal. Chem. 19:693–694, 1947.

92. Engebrigtsen, J. K. Some investigations on hypersensitiveness to bichromate in cement workers. Acta Derm.-Venereol. 32:462–468, 1952.

93. Engle, H. O., and C. D. Calnan. Chromate dermatitis from paint. Brit. J. Ind. Med. 20:192–198, 1963.

94. Englehardt, W. E., and R. L. Mayer. Über Chromekzeme im graphischen Gewerbe. Arch. Gewerbepath. Gewerbehyg. 2:140–168, 1931.

95. Environmental Protection Committee. Basic aims in environmental protection plan for Osaka area. Yosui To Haisui (Journal of Water and Waste) 12:750–758, 1970. (in Japanese)

96. Farkas, T. G., and S. L. Roberson. The effect of Cr^{3+} on the glucose utilization of isolate lenses. Exp. Eye Res. 4:124–126, 1965.

97. Federal Security Agency. Health of Workers in Chromate Producing Industry. A Study. U.S. Public Health Service Publication No. 192. Washington, D.C.: U.S. Government Printing Office, 1953. 131 pp.

98. Feldman, F. J. The state of chromium in biological materials. Fed. Proc. 27:482, 1968. (abstract)

99. Feldman, F. J., E. C. Knoblock, and W. C. Purdy. The determination of chromium in biological material by atomic absorption spectroscopy. Anal. Chim. Acta 38:489–497, 1967.

100. Feldman, F. J., and W. C. Purdy. The atomic absorption spectroscopy of chromium. Anal. Chim. Acta 33:273–278, 1965.

101. Fernley, H. N. Effects of some heavy-metal ions on purified mammalian β-glucuronidase. Biochem. J. 82:500–509, 1962.

102. Fischer, R. Die industrielle Herstellung und Verwendung der Chromverbindungen, die dabei entstehenden Gesundheitsgefahren für die Arbeiter und die Massnahmen zu ihrer Bekämpfung. Z. Gewerbe-Hyg. 18:316–319, 1911.

103. Fischer-Wasels, B. Das primäre Lungencarcinom. Acta Un. Int. Cancr. 3:140–152, 1938.

104. Fonseca, A. Dermatoses Pelo Crómio. Contribuição Para O Estudo Etiopathogénico Das Dermites De Causa Externa. Porto: Sopime Edições, 1963. 445 pp. (summaries in French, English and German)

105. Fregert, S. Book matches as a source of chromate. A.M.A. Arch. Derm. 88:546–547, 1963.

106. Fregert, S. Chromate eczema and matches. Acta Derm.-Venereol. 41:433–440, 1961.

107. Fregert, S., and P. Ovrum. Chromate in welding fumes with special reference to contact dermatitis. Acta Derma.-Venereol. 43:119–124, 1963.
108. Fregert, S., and H. Rorsman. Allergy to trivalent chromium. Arch. Derm. 90:4–6, 1964.
109. Fukai, R. Valency state of chromium in seawater. Nature 213:901, 1967.
110. Fuwa, K., W. E. C. Wacker, R. Druyan, A. F. Bartholomay, and B. L. Vallee. Nucleic acids and metals. II. Transition metals as determinants of the conformation of ribonucleic acids. Proc. Nat. Acad. Sci. U.S. 46:1298–1307, 1960.
111. Garland, T. O. Chromium plating and allied industries. Nurs. Times 38:310, 1942.
112. Glinsmann, W. H., F. J. Feldman, and W. Mertz. Plasma chromium after glucose administration. Science 152:1243–1245, 1966.
113. Glinsmann, W. H., and W. Mertz. Effect of trivalent chromium on glucose tolerance. Metabolism 15:510–520, 1966.
114. Gmelin, C. G. Experiments on the effects of baryta, strontia, chrome, molybdenum, tungsten, tellurium, titanium, osmium, platinum, iridium, rhodium, palladium, nickel, cobalt, uranium, cerium, iron and manganese, on the animal system. Edinburgh Med. Surg. J. 26:131–139, 1826.
115. Goldblatt, M. W., and V. A. J. Wagstaff. Aspects of industrial medicine and hygiene in German chemical factories. British Intelligence Objectives Subcommittee Final Report No. 1501, item 24, 1947.
116. Goldman, M., and R. H. Karotkin. Acute potassium bichromate poisoning. Amer. J. Med. Sci. 189:400–403, 1935.
117. Gooding, E., Ed. 64th Annual Report, 1954. State of Washington Department of Fisheries, 1956. 63 pp.
118. Gordon, W. A. Use of Temperature Buffered Argon Arc in Spectrographic Trace Analysis. NASA Report No. NASA-TN-D-2598. Washington, D.C.: National Aeronautics and Space Administration, 1965. 17 pp.
119. Government of India, Ministry of Labour. Chromite Section, pp. 24–50. In Investigations on the Incidence of Occupational Diseases in the Manufacture of Dichromate and in the Mining and Concentrating of Chromite. Report No. 1. New Delhi: Office of the Chief Adviser Factories, 1953. [A.M.A. Arch. Ind. Hyg. Occup. Med. 10:349–350, 1954 (abstract).]
120. Gray, S. J., and K. Sterling. The tagging of red cells and plasma proteins with radioactive chromium. J. Clin. Invest. 29:1604–1613, 1950.
121. Gresh, J. T. Chromic acid poisoning resulting from inhalation of mist developed from five per cent chromic acid solution. II. Engineering aspects of chromic acid poisoning from anodizing operations. J. Ind. Hyg. Toxicol. 26:127–130, 1944.
122. Grogan, C. H., and H. Oppenheimer. Experimental studies in metal cancerigenesis. V. Interaction of Cr(III) and Cr(VI) compounds with proteins. Arch. Biochem. Biophys. 56:204–221, 1955.
123. Gross, E. Lungenkrebs durch Arbeit in Chromat herstellenden Betrieben. (a) Technologisch-statistischer Teil, Vol. 2., pp. 966–973. In Bericht über den VIII. Internationalen Kongress für Unfallmedizin und Berufskrankheiten. Frankfort A.M., 26–28 September 1938. (2 vols.) Leipzig: Thieme, 1939.
124. Gross, E., and F. Kölsch. Über den Lungenkrebs in der Chromfarbenindustrie. Arch. Gewerbepath. Gewerbehyg. 12:164–170, 1943.

125. Gross, W. G., and V. G. Heller. Chromates in animal nutrition. J. Ind. Hyg. Toxicol. 28:52–56, 1946.
126. Gürson, C. T., and G. Saner. Effect of chromium on glucose utilization in marasmic protein–calorie malnutrition. Amer. J. Clin. Nutr. 24:1313–1319, 1971.
127. Haffty, J. Residue Method for Common Minor Elements. U.S. Geological Survey Water Supply Paper 1540-A. Washington, D.C.: U.S. Government Printing Office, 1960. 9 pp.
128. Hall, A. F. Occupational contact dermatitis among aircraft workers. J.A.M.A. 125:179–185, 1944.
129. Hambidge, K. M. Chromium nutrition in man. Presented at the 56th Annual Meeting of the Federation of American Societies for Experimental Biology, Atlantic City, April 1972.
130. Hambidge, K. M. Chromium nutrition in the mother and the growing child, pp. 169–194. In W. Mertz and W. E. Cornatzer, Eds. Newer Trace Elements in Nutrition. New York: Marcel Dekker, Inc., 1971.
131. Hambidge, K. M. Use of static argon atmosphere in emission spectrochemical determination of chromium in biological materials. Anal. Chem. 43:103–107, 1971.
132. Hambidge, K. M., B. Martinez, J. A. Jones, C. E. Boyle, W. E. Hathaway, and D. O'Brien. Correction of impaired chemotaxis of polymorphonuclear leukocytes (PMNs) from patients with diabetes mellitus by incubation with trivalent chromium in vitro. Pediatr. Res. 6:393/133, 1972. (abstract)
133. Hambidge, K. M., and D. O. Rodgerson. Comparison of the hair chromium levels of nulliparous and parous women. Amer. J. Obstet. Gynecol. 103:320–327, 1969.
134. Hanna, W. J., and C. L. Grant. Spectrochemical analysis of the foliage of certain trees and ornamentals for 23 elements. Bull. Torrey Botan. Club 89: 293–302, 1962.
135. Hanslian, L., J. Navratil, J. Jurak, and M. Kotrle. Upper respiratory tract lesions from chromic acid aerosol. Pracovni. Lekar. 19:294–298, 1967. (in Czech) (summary in English)
136. Hartelius, V. The influence of chromium on yeast spore germination and yeast growth. C. R. Trav. Lab. Carlsberg Ser. Physiol. 25:382–388, 1956.
137. Hasegawa, M. Studies on the detexicating [sic] hormone of the liver (yakriton). Effect of yakriton upon chromic nephritis. Tohoku J. Exp. Med. 32:163–176, 1938.
138. Henry, R. J., and E. C. Smith. Use of sulfuric acid–dichromate mixture in cleaning glassware. Science 104:426–427, 1946.
139. Hepler, O. E., and J. P. Simonds. Experimental nephropathies. IV. Glycosuria in dogs poisoned with uranyl nitrate, mercury bichloride and potassium dichromate. Arch. Path. 41:42–49, 1946.
140. Hermanni, F. Die Erkrankungen der in Chromatfabriken beschäftigten Arbeiter. München Med. Wchnschr. 48:536–540, 1901.
141. Herring, W. B., B. S. Leavell, L. M. Paixao, and J. H. Yoe. Trace metals in human plasma and red blood cells. A study of magnesium, chromium, nickel, copper and zinc. I. Observations of normal subjects. Amer. J. Clin. Nutr. 8:846–854, 1960.
142. Herrmann, H., and L. B. Speck. Interaction of chromate with nucleic acid in tissues. Science 119:221, 1954.

143. Hervey, R. J. Effect of chromium on the growth of unicellular Chlorophyceae and diatoms. Botan. Gaz. 111:1–11, 1949.
144. Holland, G. A., J. E. Lasater, E. D. Neumann, and W. E. Eldridge. Toxic Effects of Organic and Inorganic Pollutants on Young Salmon and Trout. State of Washington Department of Fisheries Research Bulletin No. 5, 1960. 264 pp.
145. Hopkins, L. L., Jr. Distribution in the rat of physiological amounts of injected Cr^{51}(III) with time. Amer. J. Physiol. 209:731–735, 1965.
146. Hopkins, L. L., Jr., O. Ransome-Kuti, and A. S. Majaj. Improvement of impaired carbohydrate metabolism by chromium(III) in malnourished infants. Amer. J. Clin. Nutr. 21:203–211, 1968.
147. Hopkins, L. L., Jr., and K. Schwarz. Chromium(III) binding to serum proteins, specifically siderophilin. Biochim. Biophys. Acta 90:484–491, 1964.
148. Hopps, H. C., and H. L. Cannon, Eds. Geochemical Environment in Relation to Health and Disease. Ann. N.Y. Acad. Sci. 199:1–352, 1972.
149. Horecker, B. L., E. Stotz, and T. R. Hogness. The promoting effect of chromium and the rare earths in the succinic dehydrogenase–cytochrome system. J. Biol. Chem. 128:251–256, 1939.
150. Hueper, W. C. Experimental studies in metal cancerigenesis. VII. Tissue reactions to parenterally introduced powdered metallic chromium and chromite ore. J. Nat. Cancer Inst. 16:447–469, 1955.
151. Hueper, W. C. Experimental studies in metal cancerigenesis. X. Cancerigenic effects of chromite ore roast deposited in muscle tissue and pleural cavity of rats. A.M.A. Arch. Ind. Health 18:284–291, 1958.
152. Hueper, W. C. Occupational and environmental cancers of the respiratory system. Recent Results Cancer Res. 3:1–214, 1966.
153. Hueper, W. C., and W. W. Payne. Experimental cancers in rats produced by chromium compounds and their significance to industry and public health. Amer. Ind. Hyg. Assoc. J. 20:274–280, 1959.
154. Hueper, W. C., and W. W. Payne. Experimental studies in metal carcinogenesis. Chromium, nickel, iron, arsenic. Arch. Environ. Health 5:445–462, 1962.
155. Hunter, W. C., and J. M. Roberts. Experimental study of the effects of potassium bichromate on the monkey's kidney. Amer. J. Path. 9:133–147, 1933.
156. Imbus, H. R., J. Cholak, L. H. Miller, and T. Sterling. Boron, chromium, cadmium and nickel in blood and urine. A.M.A. Arch. Environ. Health 6:286–295, 1963.
157. Ingrand, J. Biological Properties of Compounds Labeled with ^{51}Cr. Comm. Energie At. (France) Rappt. CEA-R 2585. 112 pp.
158. International Labour Office. Occupation and Health. An Encyclopaedia of Hygiene, Pathology and Social Welfare. Vol. I, pp. 444–445. Geneva: International Labour Office, 1930.
159. Ishibashi, M., and T. Skigematsu. The quantitative determination of chromium in sea water. Bull. Inst. Chem. Res. (Kyoto University) 23:59–60, 1950. (abstract)
160. Jaeger, H., and E. Pelloni. Tests épicutanes aux bichromates, positifs dans l'eczéma au ciment. Dermatologica 100:207–216, 1950.
161. Japanese Association of Industriai Health. Recommendation of Permissible Criteria of Hazardous Working Environments. Japanese Association of Industrial Health, 1969. 8 pp.
162. Joseph, K. T., V. K. Panday, S. J. Raut, and S. D. Soman. Per capita daily in-

take of trace elements from vegetables, fruits, and drinking water in India. Atom. Absorption Newslett. 7:25–27, 1968.

163. Katz, M. Quality standards for air and water. Occup. Health Rev. 17:3–8, 1965.

164. Keller, R., Ed. Basic Tables in Chemistry. New York: McGraw-Hill Book Co., Inc., 1967. 400 pp.

165. Kirchgessner, M. Wechselbeziehungen zwischen Spurenelementen in Futtermitteln und tierischen Substanzen sowie Abhängigkeitsverhältnisse zwischen einzelnen Elementen bei der Retention. II. Mitteilung Wechselbeziehungen zwischen einzelnen Spurenelementen im Wiesengras und -heu. Z. Tierphysiol. Tierern. Futtermittelkunde 14:165–175, 1959.

166. Kirchgessner, M., G. Merz, and W. Oelschlaeger. Der Einfluss des Vegetationsstadiums auf den Mengen- und Spurenelementgehalt dreier Grasarten. Arch. Tierern. 10:414–427, 1960.

167. Kleinfeld, M., and A. Rosso. Ulcerations of the nasal septum due to inhalation of chromic acid mist. Ind. Med. Surg. 34:242–243, 1965.

168. Koelsch, F. Lungenkrebs und Beruf. Acta Un. Int. Cancr. 3:243–252, 1938.

169. Koenig, P. Die Reiz- und Giftwirkungen der Chromverbindungen auf die Pflanzen. Chem.-Ztg. 35:442–443, 462–463, 1911.

170. Kondo, H., M. Kondo, K. Takeda, C. Matsuoka, T. Kitamura, and I. Tahara. On the health injury for the inhabitant in the vicinity of chromate producing factory. Ann. Rept. Tokushima Pref. Inst. Public Health 10:45–65, 1971. (in Japanese)

171. Kopp, J. F. The occurrence of trace elements in water, pp. 59–73. In D. D. Hemphill, Ed. Trace Substances in Environmental Health–III. Proceedings of University of Missouri's 3rd Annual Conference on Trace Substances in Environmental Health. Columbia: University of Missouri, 1969.

172. Koppel, J. Chromo–Natriumrhodanid. Z. Anorg. Chem. 45:359–361, 1905.

173. Koutras, G. A., M. Hattori, A. S. Schneider, F. G. Ebaugh, Jr., and W. N. Valentine. Studies on chromated erythrocytes. Effect of sodium chromate on erythrocyte glutathione reductase. J. Clin. Invest. 43:323–331, 1964.

174. Krauskopf, K. B. Factors controlling the concentrations of thirteen rare metals in sea-water. Geochim. Cosmochim. Acta 9:1–32, 1956.

175. Kubík, Š. Diseases of the respiratory tract due to dust, produced during the production of aluminum oxide. Prac. Lék. 12:458–464, 1960. (in Czech) (summary in English)

176. Kühn, K., and E. Gebhardt. Chemische und elektroneoptische Untersuchungen über die Reaktion von Chrom(III)-komplexen mit Kollagen. Z. Naturforsch. 15b:23–30, 1960.

177. Kuperman, E. F. Maximal allowable hexavalent chromium concentration in atmospheric air. In V. A. Ryazanov and M. S. Gol'dberg. Maximum Permissible Concentrations of Atmospheric Pollutants. Book 8. Moscow: "Meditsina" Press, 1964. (in Russian) (Translated by B. S. Levine in U.S.S.R. Literature on Air Pollution and Related Occupational Diseases 15:45–52, 1968.)

178. Kurylowicz, B., and S. Gasiorowski. Agrochemical study on calcinated phosphate-containing admixtures of chromium compounds. Prezemysl Chem. 37:797–800, 1958. (in Polish)

179. Kuschner, M., and S. Laskin. Experimental models in environmental carcinogenesis. Amer. J. Path. 64:183–196, 1971.

180. Langenbeck, W., M. Augustin, and C. Schäfer. Über die aktiven Metallionen des Trypsins. Hoppe Seyler Z. Physiol. Chem. 324:54–57, 1961.
181. Laskin, S., M. Kuschner, and R. T. Drew. Studies in pulmonary carcinogenesis, pp. 321–350. In M. G. Hanna, Jr., P. Nettesheim, and J. R. Gilbert, Eds. Inhalation Carcinogenesis. AEC Symposium Series 18. Oak Ridge, Tenn.: United States Atomic Energy Commission Division of Technical Information, 1970.
182. Laskin, S., L. Sohn, G. Teebor, and M. Kuschner. The experimental induction of lung cancers with chromate compounds. Amer. Ind. Hyg. Assoc. J. 29:110–111, 1968. (abstract)
183. Lederer, C. M., J. M. Hollander, and I. Perlman. Table of Isotopes, p. 18. (6th ed.) New York: John Wiley and Sons, Inc., 1967.
184. Legge, T. M. The lesions resulting from the manufacture and uses of potassium and sodium bichromate, pp. 447–454. In Oliver, T., Ed. Dangerous Trades. London: J. Murray, 1902.
185. Lehmann, K. B. Die Bedeutung der Chromate für die Gesundheit der Arbeiter. Zentralbl. Gewerbehyg. 2:193–195, 1914.
186. Lehmann, K. B. Ist Grund zu einer besonderen Beunruhigung wegen des Autftretens von Lungenkrebs bei Chromatarbeitern vorhanden? Zentralbl. Gewerbehyg. 9:168–170, 1932.
187. Letterer, E., K. Neidhardt, and H. Klett. Chromatlungenkrebs und Chromatstaublunge. Eine klinische, pathologisch-anatomische und gewerbehygienische Studie. Arch. Gewerbepath. Gewerbehyg. 12:323–361, 1944.
188. Levin, H. M., M. J. Brunner, and H. Rattner. Lithographer's dermatitis. J.A.M.A. 169:566–569, 1959.
189. Levine, R. A., D. P. H. Streeten, and R. J. Doisy. Effects of oral chromium supplementation on the glucose tolerance of elderly human subjects. Metabolism 17:114–125, 1968.
190. Loewenthal, L. J. A. Reactions in green tattoos. The significance of the valence state of chromium. A.M.A. Arch. Derm. 82:237–243, 1960.
191. Lotspeich, F. B., and E. L. Markward. Minor elements in bedrock, soil, and vegetation at an outcrop of the phosphoria formation on Snowdrift Mountain, Southeastern Idaho, pp. F1–F42. In Contributions to General Geology 1963. Geological Survey Bulletin 1181. Washington, D.C.: U.S. Government Printing Office, 1963.
192. Loveridge, B. A., G. W. C. Milner, G. A. Barnett, A. Thomas, and W. M. Henry. Determination of Copper, Chromium, Lead, and Manganese in Sea Water. Atomic Energy Research Establishment Report R-3323 (Great Britain). Harwell, England: Atomic Energy Research Establishment, 1960. 41 pp.
193. Lowman, F. G., T. R. Rice, and F. A. Richards. Accumulation and redistribution of radionuclides by marine organisms, pp. 161–199. In Radioactivity in the Marine Environment. A Report of the Panel on Radioactivity in the Marine Environment of the Committee on Oceanography, National Research Council. Washington, D.C.: National Academy of Sciences, 1971.
194. Lukanin, W. P. Zur Pathologie der Chromat-Pneumokoniose. Arch. Hyg. Bakt. 104:166–174, 1930.
195. Lumio, J. S. Lesions of the upper respiratory tract in chromium platers. Nord. Hyg. Tidskr. 5–6:86–91, 1953. (in Swedish) [A.M.A. Arch. Ind. Hyg. 8:387, 1953 (abstract).]

196. Lux, H., L. Eberle, and D. Sarre. Zur Kenntnis der Chrom(II)-Salze und des Chrom(II)-oxids. IV. Chem. Ber. 97:503–509, 1964.
197. Machle, W., and F. Gregorius. Cancer of the respiratory system in the United States chromate-producing industry. Public Health Rep. 63:(No. 35)1114–1127, Reprint No. 2882, Aug., 1948.
198. MacKenzie, R. D., R. U. Byerrum, C. F. Decker, C. A. Hoppert, and R. F. Langham. Chronic toxicity studies. II. Hexavalent and trivalent chromium administered in drinking water to rats. A.M.A. Arch. Ind. Health 18:232–234, 1958.
199. Major, R. H. Studies on a case of chromic acid nephritis. Bull. Johns Hopkins Hosp. 33:56–61, 1922.
200. Mali, J. W. H., W. J. Van Kooten, and F. C. J. Van Neer. Some aspects of the behavior of chromium compounds in the skin. J. Invest. Derm. 41:111–122, 1963.
201. Maloof, C. C. Use of edathamil calcium in treatment of chrome ulcers of the skin. A.M.A. Arch. Ind. Health 11:123–125, 1955.
202. Mancuso, T. F. Occupational cancer and other health hazards in a chromate plant. A medical appraisal. II. Clinical and toxicologic aspects. Ind. Med. Surg. 20:393–407, 1951.
203. Mancuso, T. F. Occupational cancer survey in Ohio, pp. 57–78. In Cancer Control in Public Health. Papers presented at the Sixth Annual Meeting of the Public Health Cancer Association of America, New York City, 1949.
204. Mancuso, T. F., and W. C. Hueper. Occupational cancer and other health hazards in a chromate plant. A medical appraisal. I. Lung cancers in chromate workers. Ind. Med. Surg. 20:358–363, 1951.
205. Matsuoka, C., T. Kitanura, I. Tahara, H. Kondo, and K. Takeda. A study on the measurement of the flying dust from a chromate factory. J. Jap. Soc. Air Pollut. 4:38, 1969.
206. Mazé, P., and P. J. Mazé, Jr. Influence des sels minéraux sur le pouvoir coagulant de la présure. C. R. Soc. Biol. 135:808–810, 1941.
207. Mazgon, R. Therapeutische Wirkung des Kochsalzes auf die experimentelle Chromnephritis des Kaninchens. Z. Gesamte Exp. Med. 81:195–207, 1932.
208. McCord, C. P., H. G. Higginbotham, and J. C. McGuire. Experimental chromium dermatitis. J.A.M.A. 94:1043–1044, 1930.
209. Mertz, W. Biological role of chromium. Fed. Proc. 26:186–193, 1967.
210. Mertz, W. Chromium occurrence and function in biological systems. Physiol. Rev. 49:163–239, 1969.
211. Mertz, W., and E. E. Roginski. Chromium metabolism. The glucose tolerance factor, pp. 123–153. In W. Mertz and W. E. Cornatzer, Eds. Newer Trace Elements in Nutrition. New York: Marcel Dekker, Inc., 1971.
212. Mertz, W., and E. E. Roginski. Effects of chromium(III) supplementation on growth and survival under stress in rats fed low-protein diets. J. Nutr. 97: 531–536, 1969.
213. Mertz, W., and E. E. Roginski. The effect of trivalent chromium on galactose entry in rat epididymal fat tissue. J. Biol. Chem. 238:868–872, 1963.
214. Mertz, W., E. E. Roginski, F. J. Feldman, and D. E. Thurman. Dependence of chromium transfer into the rat embryo on the chemical form. J. Nutr. 99:363–367, 1969.
215. Mertz, W., E. E. Roginski, and R. C. Reba. Biological activity and fate of trace

quantities of intravenous chromium(III) in the rat. Amer. J. Physiol. 209: 489–494, 1965.

216. Mertz, W., E. E. Roginski, and K. Schwarz. Effect of trivalent chromium complexes on glucose uptake by epididymal fat tissue of rats. J. Biol. Chem. 236: 318–322, 1961.

217. Mertz, W., and K. Schwarz. Impaired intravenous glucose tolerance as an early sign of dietary necrotic liver degeneration. Arch. Biochem. Biophys. 58:504–506, 1955.

218. Mertz, W., and K. Schwarz. Relation of glucose tolerance factor to impaired glucose tolerance in rats on stock diets. Amer. J. Physiol. 196:614–618, 1959.

219. Meyers, J. B. Acute pulmonary complications following inhalation of chromic acid mist. Preliminary observations of two patients who inhaled massive amounts of chromic acid. A.M.A. Arch. Ind. Hyg. Occup. Med. 2:742–747, 1950.

220. Michael, S. E. The precipitation of proteins with complex salts. Biochem. J. 33:924–930, 1939.

221. Mikosha, A. Trace elements in human embryos. Nauk. Zap. Stanislavs'k Med. Inst. (3):85–89, 1959. (in Ukranian) (Chem. Abstr. 59:7969, 1963)

222. Milby, T. H., M. M. Key, R. L. Gibson, and H. E. Stokinger. Chemical hazards, pp. 63–242. In W. M. Gafafer, Ed. Occupational Diseases. A Guide to Their Recognition. U.S. Public Health Service Publication 1097. Washington, D.C.: U.S. Government Printing Office, 1966.

223. Morris, G. E. Chromate dermatitis from chrome glue and other aspects of the chrome problem. A.M.A. Arch. Ind. Health 11:368–371, 1955.

224. Morris, G. E. "Chrome" dermatitis. A study of the chemistry of shoe leather with particular reference to basic chromic sulfate. A.M.A. Arch. Derm. 78: 612–618, 1958.

225. Moshier, R. W., and R. E. Sievers. Gas Chromatography of Metal Chelates. New York: Pergamon Press, Inc., 1965. 163 pp.

226. Murakami, Y., Y. Suzuki, T. Yamagata, and N. Yamagata. Chromium and manganese in Japanese diet. J. Radiat. Res. 6:105–110, 1965.

227. Naismith, W. E. F. The cross-linking of conarachin II with metal salts. Arch. Biochem. Biophys. 73:255–261, 1958.

228. National Academy of Sciences. Water Quality Criteria, 1972. Washington, D.C.: U.S. Government Printing Office. (in press)

229. National Research Council. Trends in Usage of Chromium. Report of the Panel on Chromium of the Committee on Technical Aspects of Critical and Strategic Materials. National Materials Advisory Board Report NMAB-256. Washington, D.C.: National Academy of Sciences, 1970. 88 pp.

230. Nettesheim, P., M. G. Hanna, Jr., D. G. Doherty, R. F. Newell, and A. Hellman. Effect of calcium chromate dust, influenza virus and 100 R whole-body x-radiation on lung tumor incidence in mice. J. Nat. Cancer Inst. 47: 1129–1144, 1971.

231. Nettesheim, P., M. G. Hanna, Jr., D. G. Doherty, R. F. Newell, and A. Hellman. Effects of chronic exposure to artificial smog and chromiun oxide dust on the incidence of lung tumors in mice, pp. 305–320. In M. G. Hanna, Jr., P. Nettesheim, and J. R. Gilbert, Eds. Inhalation Carcinogenesis. AEC Symposium Series 18. Oak Ridge, Tenn.: United States Atomic Energy Commission Division of Technical Information, 1970.

232. Newhouse, M. L. A cause of chromate dermatitis among assemblers in an automobile factory. Brit. J. Ind. Med. 20:199–203, 1963.
233. Ohta, F. Studies on the detoxicating hormone of the liver (yakriton). Contribution to the usage of yakriton against experimental chromate nephritis. Tohoku J. Exp. Med. 39:37–46, 1940.
234. Olson, P. A. Comparative toxicity of Cr(VI) and Cr(III) in salmon, pp. 215–218. In Hanford Biological Research Annual Report for 1957. HW-53500. Richland, Washington, 1958.
235. Olson, P. A., and R. F. Foster. Effect of chronic exposure to sodium dichromate on young chinook salmon and rainbow trout, pp. 35–47. In Hanford Biological Research Annual Report for 1955. HW-41500. Richland, Washington, 1956.
236. Olson, P. A., and R. F. Foster. Further studies on the effect of sodium dichromate on juvenile chinook salmon, pp. 214–224. In Hanford Biological Research Annual Report for 1956. HW-45700. Richland, Washington, 1957.
237. Ophüls, W. Experimental nephritis in guinea-pigs by subcutaneous injections of chromates. Proc. Soc. Exp. Biol. Med. 9:11–12, 1911.
238. Ophüls, W. Experimental nephritis in rabbits by subcutaneous injections of chromates. Proc. Soc. Exp. Biol. Med. 9:13, 1911.
239. Parkhurst, H. J. Dermatosis industrialis in a blue print worker due to chromium compounds. A.M.A. Arch. Derm. Syph. 12:253–256, 1925.
240. Partington, C. N. Acute poisoning with potassium bichromate. Brit. Med. J. 2:1097–1098, 1950.
241. Payne, W. W. Production of cancers in mice and rats by chromium compounds. A.M.A. Arch. Ind. Health 21:530–535, 1960.
242. Payne, W. W. The role of roasted chromite ore in the production of cancer. Arch. Environ. Health 1:20–26, 1960.
243. Pchelin, V. A., N. V. Grigor'eva, and V. N. Izmailova. Fixation of polypeptide chains in two conformations. Dokl. Akad. Nauk. SSSR 151:(1)134–135, 1963. (in Russian) (Chem. Abstr. 59:10229g, 1963)
244. Pekarek, R. S., and E. C. Hauer. Direct determination of serum chromium and nickel by an atomic absorption spectrophotometer with a heated graphite furnace. Fed. Proc. 31:700, 1972. (abstract)
245. Perlman, D. Some effects of metallic ions on the metabolism of *Aerobacter aerogenes*. J. Bact. 49:167–175, 1945.
246. Pfeil, E. Lungentumoren als Berufsenkrankung in Chromatbetrieben. Deutsch. Med. Wochenschr. 61:1197–1200, 1935.
247. Pickering, Q. H., and C. Henderson. The acute toxicity of some heavy metals to different species of warmwater fishes. Air Water Pollut. 10:453–463, 1966.
248. Pickett, E. E. Current capabilities in analysis of trace substances. Flame photometry and atomic absorption, pp. 29–36. In D. D. Hemphill, Ed. Proceedings of University of Missouri's 1st Annual Conference on Trace Substances in Environmental Health. Columbia: University of Missouri, 1967.
249. Pierce, J. O., II, and J. Cholak. Lead, chromium and molybdenum by atomic absorption. A.M.A. Arch. Environ. Health 13:208–212, 1966.
250. Pierce, J. O., and L. D. Scheel. Toxicity of alloys of chromium. I. Solubility of chromium in solutions of water and protein. Arch. Environ. Health 10: 870–876, 1965.
251. Pinta, M. Detection and Determination of Trace Elements. Absorption, Spec-

trophotometry, Emission, Spectroscopy, Polarography. (3rd ed.) Ann Arbor: Ann Arbor Science Publishers, 1971. 588 pp.

252. Pirilä, V., and O. Kilpiö. On dermatoses caused by bichromates. Acta Derm.-Venereol. 29:550–563, 1949.

253. Pirozzi, D. J., P. R. Gross, and M. H. Samitz. The effect of ascorbic acid on chrome ulcers in guinea pigs. Arch. Environ. Health 17:178–180, 1968.

254. Popova, Ya., S. Chaga, and M. Beshkova. Éffect of certain elements on the biosynthesis of vitamin B_{12} from Propionibacterium shermanii. Nauchni Tr., Nauchnoizzled. Inst. Konservna Prom., Plovidiv 1:147–151, 1963. (in Russian) (Chem. Abstr. 61:7656b, 1964)

255. Pratt, P. F. Chromium, pp. 136–141. In H. D. Chapman, Ed. Diagnostic Criteria for Plants and Soils. Riverside: University of California Division of Agricultural Sciences, 1966.

256. Priestley, J. Observations on the physiological action of chromium. J. Anat. Physiol. 11:285–301, 1877.

257. Princi, F., L. H. Miller, A. Davis, and J. Cholak. Pulmonary disease of ferro-alloy workers. J. Occup. Med. 4:301–310, 1962.

258. Pringle, B. H., D. E. Hissong, E. L. Katz, and S. T. Mulawka. Trace metal accumulation by estuarine mollusks. J. Sanit. Eng. Div., Proceedings of the American Society of Civil Engineers. Proceedings Paper 5970, June 1968. 94(SA3), pp. 455–475 (Eng.).

259. Public Health Service drinking water standards. Public Health Rep. 61:371–384, 1946.

260. Puck, T. T., A. Garen, and J. Cline. The mechanism of virus attachment to host cells. I. The role of ions in the primary reaction. J. Exp. Med. 93:65–88, 1951.

261. Rambousek, J. Industrial Poisoning from Fumes, Gases and Poisons of Manufacturing Processes. Translated and edited by T. M. Legge. London: E. Arnold, 1913. 360 pp.

262. Raymont, J. E. G., and J. Shields. Toxicity of copper and chromium in the marine environment. Int. J. Air Water Pollut. 7:435–443, 1963.

263. Raymont, J. E. G., and J. Shields. Toxicity of copper and chromium in the marine environment, pp. 275–283. In E. A. Pearson, Ed. Advances in Water Pollution Research. Proceedings of the International Conference held in London, September, 1962. Vol. 3. New York: The Macmillan Company, 1964.

264. Regan, T. M., and M. M. Peters. Heavy metals in digesters. Failure and cure. J. Water Pollut. Control Fed. 42:1832–1839, 1970.

265. Remy, H. Treatise on Inorganic Chemistry. 2 vols. New York: Elsevier Publishing Co., 1956. 1,666 pp.

266. Rjazanov, V. A. Criteria and methods for establishing maximum permissible concentration of air pollution. Bull. WHO 32:389–398, 1965.

267. Robinson, W. O. The Inorganic Composition of Some Important American Soils. U.S. Department of Agriculture Bulletin No. 122. Washington, D.C.: U.S. Department of Agriculture, 1914. 27 pp.

268. Robinson, W. O., G. Edgington, and H. G. Byers. Chemical Studies of Infertile Soils Derived from Rocks High in Magnesium and Generally High in Chromium and Nickel. U.S. Department of Agriculture Technical Bulletin No. 471. Washington, D.C.: U.S. Department of Agriculture, 1935. 28 pp.

269. Roddy, W. T., and R. M. Lollar. Resistance of white leather to breakdown by perspiration. J. Amer. Leather Chemists Assoc. 50:180–192, 1955.

270. Roe, F. J. C., and R. L. Carter. Chromium carcinogenesis. Calcium chromate as a potent carcinogen for the subcutaneous tissues of the rat. Brit. J. Cancer 23:172–176, 1969.

271. Roginski, E. E., and W. Mertz. An eye lesion in rats fed low-chromium diets. J. Nutr. 93:249–251, 1967.

272. Roginski, E. E., and W. Mertz. Effects of chromium(III) supplementation on glucose and amino acid metabolism in rats fed a low protein diet. J. Nutr. 97:525–530, 1969.

273. Rollinson, C. L. Problems of chromium reactions, pp. 429–446. In G. A. Andrews, R. M. Kniseley, and H. N. Wagner, Jr., Eds. Radioactive Pharmaceuticals. U.S. Atomic Energy Commission CONF-65111. Oak Ridge, Tenn.: U.S. Atomic Energy Commission, 1966.

274. Rollinson, C. L. Chromium, molybdenum and tungsten, pp. 623–769. In Comprehensive Inorganic Chemistry. Oxford: Pergamon Press, 1973.

275. Romoser, G. L., W. A. Dudley, L. J. Machlin, and L. Loveless. Toxicity of vanadium and chromium for the growing chick. Poultry Sci. 40:1171–1173, 1961.

276. Ruhling, A. Heavy Metals within the Region of Vargo-Trollhattan. Lund University (Sweden) Institute of Ecological Botany Report 14. April 1970. 22 pp. (in Swedish)

277. Samitz, M. H. Recent developments in industrial dermatology. Presented at the 161st Annual Meeting of the Connecticut State Medical Society, April 29, 1953.

278. Samitz, M. H., and S. Gross. Effects of hexavalent and trivalent chromium compounds on the skin. A.M.A. Arch. Derm. 84:404–409, 1961.

279. Samitz, M. H., and S. Gross. Extraction by sweat of chromium from chrome tanned leathers. J. Occup. Med. 2:12–14, 1960.

280. Samitz, M. H., S. Gross, and S. Katz. Inactivation of chromium ion in allergic eczematous dermatitis. J. Invest. Derm. 38:5–12, 1962.

281. Samitz, M. H., and S. Katz. A study of the chemical reactions between chromium and skin. J. Invest. Derm. 43:35–43, 1964.

282. Samitz, M. H., and S. Katz. Preliminary studies on the reduction and binding of chromium with skin. Arch. Derm. 88:816–819, 1963.

283. Samitz, M. H., and S. Katz. Protection against inhalation of chromic acid mist. Use of filters impregnated with ascorbic acid. Arch. Environ. Health 11:770–772, 1965.

284. Samitz, M. H., S. Katz, and S. Gross. Nature of the chromium extracted from leather by sweat. J. Occup. Med. 2:435–436, 1960.

285. Samitz, M. H., D. M. Scheiner, and S. A. Katz. Ascorbic acid in the prevention of chrome dermatitis. Mechanism of inactivation of chromium. Arch. Environ. Health 17:44–45, 1968.

286. Samitz, M. H., and J. Shrager. Prevention of dermatitis in the printing and lithographing industries. A.M.A. Arch. Derm. 94:307–309, 1966.

287. Samitz, M. H., J. Shrager, and S. Katz. Studies on the prevention of injurious effects of chromates in industry. Ind. Med. Surg. 31:427–432, 1962.

288. Samsonov, G. V., Ed. Handbook of the Physicochemical Properties of the Elements. New York: IFI/Plenum, 1968. 941 pp.

289. Savory, J., P. Mushak, N. O. Roszel, and F. W. Sunderman, Jr. Determination of chromium in serum by gas chromatography. Fed. Proc. 27:777, 1968. (abstract)
290. Schinz, H. R. Der Metallkrebs. Ein neues Prinzip der Krebserzeugung. Schweiz. Med. Wchnschr. 72:1070–1074, 1942.
291. Schinz, H. R., and E. Uehlinger. Der Metallkrebs. Ein neues Prinzip der Krebserzeugung. Z. Krebsforsch. 52:425–437, 1942.
292. Schroeder, H. A. Chromium. Air Quality Monograph #70-15. Washington, D.C.: American Petroleum Institute, 1970. 28 pp.
293. Schroeder, H. A. Chromium deficiency in rats. A syndrome simulating diabetes mellitus with retarded growth. J. Nutr. 88:439–445, 1966.
294. Schroeder, H. A. The biological trace elements or peripatetics through the periodic table. J. Chron. Dis. 18:217–228, 1965. (editorial)
295. Schroeder, H. A. The role of chromium in mammalian nutrition. Amer. J. Clin. Nutr. 21:230–244, 1968.
296. Schroeder, H. A., and J. J. Balassa. Influence of chromium, cadmium, and lead on rat aortic lipids and circulating cholesterol. Amer. J. Physiol. 209: 433–437, 1965.
297. Schroeder, H. A., J. J. Balassa, and I. H. Tipton. Abnormal trace metals in man—chromium. J. Chron. Dis. 15:941–964, 1962.
298. Schroeder, H. A., J. J. Balassa, and W. H. Vinton, Jr. Chromium, cadmium and lead in rats. Effects on life span, tumors and tissue levels. J. Nutr. 86: 51–66, 1965.
299. Schroeder, H. A., J. J. Balassa, and W. H. Vinton, Jr. Chromium, lead, cadmium, nickel and titanium in mice. Effect on mortality, tumors and tissue levels. J. Nutr. 83:239–250, 1964.
300. Schroeder, H. A., A. P. Nason, and I. H. Tipton. Chromium deficiency as a factor in atherosclerosis. J. Chron. Dis. 23:123–142, 1970.
301. Schulz, H. Ueber Hefegifte. Pfluegers Arch. Ges. Physiol. 42:517–541, 1888.
302. Schwartz, L., L. Tulipan, and D. J. Birmingham. Occupational Diseases of the Skin. (3rd ed.) Philadelphia: Lea and Febiger, 1957. 981 pp.
303. Schwarz, K., and W. Mertz. Chromium(III) and the glucose tolerance factor. Arch. Biochem. Biophys. 85:292–295, 1959. (letter to the editor)
304. Shacklette, H. T., J. C. Hamilton, J. G. Boerngen, and J. M. Bowles. Elemental Composition of Surficial Materials in the Conterminous United States. U.S. Geological Survey Paper 574-D. Washington, D.C.: U.S. Government Printing Office, 1971. 71 pp.
305. Shelley, W. B. Chromium in welding fumes as cause of eczematous hand eruption. J.A.M.A. 189:772–773, 1964.
306. Shepherd, C. M., and R. L. Jones. Hexavalent Chromium. Toxicological Effects and Means for Removal from Aqueous Solution. NRL Report 7215. Washington, D.C.: Naval Research Laboratory, 1971. 18 pp.
307. Sherman, L., J. A. Glennon, W. J. Brech, G. H. Klomberg, and E. S. Gordon. Failure of trivalent chromium to improve hyperglycemia in diabetes mellitus. Metabolism 17:439–442, 1968.
308. Shimkin, M. M., and J. Leiter. Induced pulmonary tumors in mice. III. Role of chronic irritation in the production of pulmonary tumors in strain A mice. J. Nat. Cancer Inst. 1:241–254, 1940.

309. Sievers, R. E. Gas chromatographic and related studies of metal complexes, pp. 270–288. In S. Kirschner, Ed. Coordination Chemistry. Papers Presented in Honor of John C. Bailar, Jr. New York: Plenum Press, 1969.
310. Sievers, R. E., J. W. Connolly, and W. D. Ross. Metal analysis by gas chromatography of chelates of heptafluorodimethyloctanedione. J. Gas Chromatogr. 5:241–247, 1967.
311. Silverman, L., and J. F. Ege, Jr. A rapid method for determination of chromic acid mist in air. J. Ind. Hyg. Toxicol. 29:136–139, 1947.
312. Silverman, L., and J. F. Ege, Jr. Chromium compounds in gaseous atmospheres. U.S. Patent No. 2,483,108, September 27, 1949.
313. Simavoryan, P. S. Effect of sexivalent chromium on concentration and dilution capacity of dog kidney. Chem. Abstr. 66:74481s, 1967.
314. Simonds, J. P., and O. E. Hepler. Experimental nephropathies. I. A method of producing controlled selective injury of renal units by means of chemical agents. Arch. Path. 39:103–108, 1945.
315. Simons, E. N. Guide to Uncommon Metals. London: F. Muller, 1967. 245 pp.
316. Skog, E. Positive patch test to trivalent chromium. Acta Derm.-Venereol. 35:393, 1955.
317. Smith, A. R. Chrome poisoning with manifestations of sensitization. Report of a case. J.A.M.A. 97:95–98, 1931.
318. Smithells, C. J., Ed. Metals Reference Book. Vol. I. (4th ed.) New York: Plenum Press, 1967. 370 pp.
319. Sollmann, T. A Manual of Pharmacology and Its Applications to Therapeutics and Toxicology, p. 996. (5th ed.) Philadelphia: W. B. Saunders Co., 1936.
320. Spier, H. W., R. Natzel, and G. Pascher. Das Chromatekzem (mit besonderer Berücksichtigung der ätiologischen Bedeutung des Spurenchromates). Arch. Gewerbepath. Gewerbehyg. 14:373–407, 1956.
321. State of California Department of Water Resources. Investigation of Geothermal Waters in the Long Valley Area, Mono County. 1967. 141 pp.
322. Steffee, C. H., and A. M. Baetjer. Histopathologic effects of chromate chemicals. Report of studies in rabbits, guinea pigs, rats, and mice. Arch. Environ. Health 11:66–75, 1965.
323. Stiasny, E. Gerbereichemie (Chromgerbung). Dresden: Steinkopff, 1931. 586 pp.
324. Stickland, L. H. The activation of phosphoglucomutase by metal ions. Biochem. J. 44:190–197, 1949.
325. Stokinger, H. E. Current problems of setting occupational exposure standards. Arch. Environ. Health 19:277–281, 1969.
326. Streeten, D. H. P., M. M. Gerstein, B. M. Marmor, and R. J. Doisy. Reduced glucose tolerance in elderly human subjects. Diabetes 14:579–583, 1965.
327. Sullivan, R. J. Preliminary Air Pollution Survey of Chromium and its Compounds. A Literature Review. NAPCA Publication APTD 69-34. Raleigh: National Air Pollution Control Administration, 1969. 75 pp.
328. Taylor, F. H. The relationship of mortality and duration of employment as reflected by a cohort of chromate workers. Amer. J. Public Health 56:218–229, 1966.
329. Teleky, L. Krebs bei Chromarbeitern. Deutsch Med. Wochenschr. 62:1353, 1936.
330. Tenny, A. M., and G. H. Stanley. Application of atomic absorption spectros-

copy for monitoring selected metals in an industrial waste. Purdue Univ. Eng. Ext. Series #129, 455–467, 1967.

331. The Merck Index of Chemicals and Drugs. (7th ed.) Rahway, N.J.: Merck and Co., Inc., 1960. 1,639 pp.

332. Tipton, I. H. Distribution of trace metals in the human body, pp. 27–42. In M. J. Seven, Ed. Metal Binding in Medicine. Philadelphia: J. B. Lippincott Co., 1960.

333. Tipton, I. H., P. L. Stewart, and J. Dickson. Patterns of elemental excretion in long term balance studies. Health Phys. 16:455–462, 1969.

334. Tipton, I. H., P. L. Stewart, and P. G. Martin. Trace elements in diets and excreta. Health Phys. 12:1683–1689, 1966.

335. Toepfer, E. W., W. Mertz, E. E. Roginski, and M. M. Polansky. Chromium in foods in relation to biological activity. J. Agric. Food Chem. 21:69–73, 1973.

336. Trumper, H. B. The health of the worker in chromium plating. Brit. Med. J. 1:705–706, 1931.

337. Tuchkova, T. G. Effect of chromium salts on the quality of mulberry silkworm cocoons. Tr. Turkmensk Sel'skokhoz. Inst. 11(1):61–63, 1962. (in Russian) (Chem. Abstracts 63:8794a, 1965)

338. Udy, M. J. History of chromium, pp. 1–13. In M. J. Udy, Ed. Chromium. Vol. I. Chemistry of Chromium and Its Compounds. American Chemical Society Monograph #132. New York: Reinhold Publishing Corporation, 1956.

339. United Nations, Food and Agriculture Organization, Department of Fisheries, Fishery Resources Division. FAO Fisheries Report No. 99, Suppl. 1. Report of the Seminar on Methods of Detection, Measurement and Monitoring of Pollutants in the Marine Environment, Rome, 4–10 December 1970, Supplement to the Report of the Technical Conference on Marine Pollution and its Effects on Living Resources and Fishing. Rome: United Nations, 1971. 123 pp.

340. U.S. Department of Health, Education, and Welfare, National Air Pollution Control Administration. Air Quality Data from the National Air Sampling Networks and Contributing State and Local Networks. 1966 Edition, p. 88. Publication APTD 68-9. Washington, D.C.: U.S. Government Printing Office, 1968.

341. U.S. Department of Health, Education, and Welfare, Public Health Service. Interaction of Heavy Metals and Biological Sewage Treatment Processes. Environmental Health Series, Water Supply and Pollution Control. PHS Publication 999-WP-22. Cincinnati: Public Health Service, 1965. 193 pp.

342. U.S. Department of Health, Education, and Welfare, Public Health Service, Consumer Protection and Environmental Health Services, Environmental Control Administration. Public Health Service Drinking Water Standards (rev. 1962). PHS Publication 956. Washington, D.C.: U.S. Government Printing Office, 1962. 61 pp.

343. U.S. Department of Health, Education, and Welfare, Public Health Service, Division of Air Pollution. Air Quality Data from the National Air Sampling Networks and Contributing State and Local Networks. 1964–1965 Edition. Cincinnati: U.S. Department of Health, Education, and Welfare, 1966. 106 pp.

344. U.S. Department of Health, Education, and Welfare, Public Health Service, Division of Air Pollution. Air Quality Data of the National Air Sampling

Networks, 1962. Cincinnati: U.S. Department of Health, Education, and Welfare, 1964. 50 pp.

345. U.S. Department of Health, Education, and Welfare, Public Health Service, Division of Air Pollution, Robert A. Taft Sanitary Engineering Center, Cincinnati. Air Pollution Measurements of the National Air Sampling Networks; Analyses of Suspended Particulates, 1963. Washington, D.C.: U.S. Government Printing Office, 1965. 87 pp.

346. U.S. Department of Health, Education, and Welfare, Public Health Service, Environmental Health Service, Bureau of Water Hygiene. Community Water Supply Study. Analysis of National Survey Findings. Washington, D.C.: U.S. Department of Health, Education, and Welfare, 1970. 111 pp.

347. U.S. Department of Labor. Occupational Safety and Health Standards, Chap. XVII, Part 1910.93. Air Contaminants, Tables G1 and G2. Pub. Fed. Register, Vol. 36, No. 157, Aug. 13, 1971, p. 15101.

348. U.S. Environmental Protection Agency. Air Quality Data for 1967 from the National Air Surveillance Networks and Contributing State and Local Networks (rev. 1971), p. 101. Publication APTD 0741. Research Triangle Park, N.C.: U.S. Environmental Protection Agency, 1971.

349. U.S. Environmental Protection Agency. Air Quality Data for 1968 from the National Air Surveillance Networks and Contributing State and Local Networks, p. 108. Publication APTD-0978. Washington, D.C.: U.S. Government Printing Office, 1972.

350. U.S. Environmental Protection Agency, Water Quality Office, Analytical Quality Control Laboratory. Methods for Chemical Analysis of water and Wastes, 1971. Washington, D.C.: U.S. Government Printing Office, 1971. 312 pp.

351. Underwood, E. J. Trace Elements in Human and Animal Nutrition. (3rd ed.) New York: Academic Press, Inc., 1971. 543 pp.

352. Urone, P. F. Stability of colorimetric reagent for chromium, s-diphenylcarbazide, in various solvents. Anal. Chem. 27:1354–1355, 1955.

353. van der Walt, C. F. J., and A. J. van der Merwe. Colorimetric determination of chromium in plant ash, soil, water and rocks. Analyst 63:809–811, 1938.

354. Vigliani, E. C., and N. Zurlo. Erfahrungen der Clinica del Lavoro mit einigen maximalen Arbeitsplatzkonzentrationen (MAK) von Industriegiften. Arch. Gewerbepath. Gewerbehyg. 13:528–534, 1955.

355. Visek, W. J., I. B. Whitney, U. S. G. Kuhn, III, and C. L. Comar. Metabolism of Cr^{51} by animals as influenced by chemical state. Proc. Soc. Exp. Biol. Med. 84:610–615, 1953.

356. Völkl, A. Tages-chromausscheidung von Normalpersonen. Zentralbl. Arbeitsmed. 21:122, 1971.

357. Wacker, W. E. C., and B. L. Vallee. Chromium, manganese, nickel and other metals in RNA. Fed. Proc. 18:345, 1959. (abstract).

358. Wacker, W. E. C., and B. L. Vallee. Nucleic acids and metals. I. Chromium, manganese, nickel, iron and other metals in ribonucleic acid from diverse biological sources. J. Biol. Chem. 234:3257–3262, 1959.

359. Walsh, E. N. Chromate hazards in industry. J.A.M.A. 153:1305–1308, 1953.

360. Weeks, M. E., and H. M. Leicester. Discovery of the Elements, p. 279. (7th ed.) Easton, Pa.: Mack Printing Co., 1968.

361. Weinberg, E. D. Manganese requirement for sporulation and other secondary biosynthetic processes of Bacillus. Appl. Microbiol. 12:436–441, 1964.

362. West, P. W. Chemical analysis of inorganic particulate pollutants, pp. 147–185. In A. C. Stern, Ed. Air Pollution. Vol. II. Analysis, Monitoring and Surveying. (2nd ed.) New York: Academic Press, 1968.
363. White, R. P. The Dermatergoses or Occupational Affections of the Skin. (4th ed.) London: H. K. Lewis and Co., Ltd., 1934. 716 pp.
364. Winston, J. R., and E. N. Walsh. Chromate dermatitis in railroad employees working with diesel locomotives. J.A.M.A. 147:1133–1134, 1951.
365. Wöhlbier, W., M. Kirchgessner, and W. Oelschläger. Der Gehalt an Mengen- und Spurenelementen in verschieden Kartoffel- und Rübensorten. Z. Tierern. Futtermittelk. 12:259–262, 1957.
366. Wöhlbier, W., M. Kirchgessner, and W. Oelschläger. Die Gehalte des Rotklees und der Luzerne an Mengen- und Spurenelementen. Arch. Tierern. 9:194–201, 1959.
367. Wolstenholme, W. A. Analysis of dried blood plasma by spark source mass spectrometry. Nature 203:1284–1285, 1964.
368. World Health Organization. European Standards for Drinking Water, p. 33. (2nd ed.) Geneva: World Health Organization, 1970.
369. World Health Organization. Manual of the International Statistical Classification of Diseases, Injuries, and Causes of Death. (7th rev., 2 vols.) Geneva: World Health Organization, 1957. 933 pp.
370. Wutzdorff, R. Die in Chromatfabriken beobachteten Gesundheitsschädigungen und die zur Verhütang derselben erforderlichen Massnahmen. Arbeiten aus dem kaiserlichen Gesundheitsamte 13:328–335, 1896–7.
371. Yanin, L. V., Ed. Labor Hygiene and Industrial Sanitation. A Collection of the Principal Official Publications. Part I. (translated from Russian) Jerusalem: S. Monson, 1962. 280 pp.
372. Zina, G. Cromoreattività cutanea e dermatosi professionali. Minerva Derm. 31:305–318, 1956.
373. Zvaifler, N. I. Chromic acid poisoning resulting from inhalation of mist developed from five per cent chromic acid solution. I. Medical aspects of chromic acid poisoning. J. Ind. Hyg. Toxicol. 26:124–126, 1944.

Index